Dream Machines

Technographies

Series Editors: Steven Connor, David Trotter and James Purdon

How was it that technology and writing came to inform each other so extensively that today there is only information? Technographies seeks to answer that question by putting the emphasis on writing as an answer to the large question of 'through what?'. Writing about technographies in history, our contributors will themselves write technographically.

Dream Machines

Steven Connor

OPEN HUMANITIES PRESS
London 2017

First edition published by Open Humanities Press 2017

Copyright © 2017 Steven Connor

This is an open access book, licensed under Creative Commons By Attribution Share Alike license. Under this license, authors allow anyone to download, reuse, reprint, modify, distribute, and/or copy their work so long as the authors and source are cited and resulting derivative works are licensed under the same or similar license. No permission is required from the authors or the publisher. Statutory fair use and other rights are in no way affected by the above. Read more about the license at creativecommons.org/licenses/by-sa/4.0

Freely available online at:
http://www.openhumanitiespress.org/books/titles/dream-machines/

Cover Art, figures, and other media included with this book may be under different copyright restrictions.

Cover Illustration: Writin, Machin, Cod (2016) gouache on paper
Copyright © 2016 Navine G. Khan-Dossos, Creative Commons CC-BY-NC-ND

Print ISBN 978-1-78542-036-8
PDF ISBN   978-1-78542-037-5

OPEN HUMANITIES PRESS

Open Humanities Press is an international, scholar-led open access publishing collective whose mission is to make leading works of contemporary critical thought freely available worldwide. More at http://openhumanitiespress.org

Contents

1. Psychotechnographies   7
2. Transports   25
3. Dream Machines   50
4. Pleasure Machines   66
5. Medical Machines   103
6. Radiation Machines   118
7. Invisibility Machines   143
8. Perpetual Motion   160
9. Shutdown   186
    Works Cited   189
    Index   207

I

Psychotechnographies

the soul it self by these middle things recollecteth it self

Heinrich Cornelius Agrippa,
Three Books of Occult Philosophy (1651)

Machines are thought to belong to the realm of the definite and the actual, and therefore to be opposed to the softer, less substantial world of dream, fantasy and vision. The aim of this book is to use the analysis of a number of imaginary or frankly impossible machines to isolate a certain strain of the visionary that may be involved in all thinking and writing about machines. Technologies are 'dreamed up' and machines require a great deal of lucid dreamwork in order to be brought into being. So the actual development of technologies like those of flight or phonography has usually been preceded by centuries of more or less concrete embodiment in fantasy. This means that the very idea of the machine becomes entangled with magical thinking. Sometimes these dreams are extreme wish-fulfilment, the ultimate cognitive labour-saving device; but they can also be nightmare visions, machines that have no other purpose than to propagate or perpetuate themselves, like Novalis's 1799 vision of 'a monstrous mill, driven by the stream of chance and floating on it, a mill of itself without builder or miller and really a true *perpetuum mobile*, a mill grinding itself' (Novalis 1997: 144), or Conrad's vision, almost a century later, of the 'knitting-machine' of the cosmos, that 'has made itself without thought, without conscience, without foresight, without eyes, without heart [...] It knits us in and it knits us out' (Conrad 1983: 425). Machines that depend on or result from dream have a way of turning back on the dreamer, and figuring the machinery of the act of dreaming itself. It is this apparitional function of the dream machine that much literary writing isolates and allows us to think about. Machines are not real; they are machines of the real. This book aims to be a contribution to technography, which may be defined at its simplest through the spelling out of its elements: technography is the writing of technology. In its earliest uses, in classical and Hellenistic writing, the word referred to a technical treatise on rhetoric, with Aristotle's *Rhetoric to Alexander* and *Rhetoric* being described as the two earliest examples (Kraus

2011: 279). This is still the sense in which the word 'technography' is employed by historians of rhetoric (Heath 2004). 'Technography' has also been used to mean the technical language employed by a particular profession, as in a 1920 reference in a discussion of silver candlesticks in the *Burlington Magazine* to 'silver mountings as they are called in the technography of the trade' (Veitch 1920: 18). In 1963, H.C. McDaniel, an ex-president of the Society for Technical Writers and Professionals, proposed the word technography as 'the generic descriptor encompassing all the functions normally associated with technical writing, editing, illustrating, publishing and allied operations' (Galasso 1963: 23). The *SWTP Review*, the house journal of the Society of Technical Writers and Professionals was accordingly subtitled *The Journal of Technography* for some years.

Douglas Kahn's deployment of the term (2004) to describe his study of the use of technology in the work of composer John Bischoff marks a shift to the idea of technography as an historical account of a technology. In 2007, Greenwood Press released fourteen volumes of what it called Technographies, each subtitled *The Life Story of a Technology*. The series, which included volumes on the railway, firearms, the car, the radio, and the book aimed to provide something like a biography, each volume outlining 'the artifact's antecedents or "family" background, its "youthful" development, subsequent maturation, and either its eventual absorption into ubiquitous societal adoption or its decline and obsolescence' (Cutliffe 2007: 165). The term 'technography has found a variety of further uses, in fields as diverse as ethnography (Woolgar 1998: 444; Kien 2008: 1101), techno-culture (Vannini, Hodson and Vannini 2009: 462; Jansen and Vellema 2011: 169), computer-mediated collective writing (Ball n.d.), and the ludic experiments of Oulipo (Harris 1970).

Technography therefore seems to involve a kind of oscillation of positions, not to say specifically of prepositions, with regard to the technical, in which writing *of* techne can switch into the writing *in* or *through* techne. James Purdon derives his notion of technography from 'a union of the material and the symbolic [which] is ingrained within the concept of technology itself from the earliest times' (Purdon 2016). His interest, like mine, is in the ways in which discourse attaches to machines and mechanisms, and 'a specific kind of discourse which prioritizes an awareness of the process of its own mediation; a kind of writing which seeks to bring to consciousness the technicity of text and the textuality of technics; the kind of writing which we have sometimes been inclined to call literature, but which encompasses symbolic operations – techniques or indeed technics – across many media forms' (Purdon 2016). My own definition of technography, as 'any writing about any technology that implicates or is attuned to the technological condition of its own writing' is in accord with this. Technography is not just one mode or mood of literary writing; rather, all the writing we tend to call literary must be regarded as technographic, in that it is 'the engineering in writing of the particular kind of engine of writing it aims at being. So modern literary writing is ever more technographic, not in the simple sense that it is concerned with other kinds of machinery, but in the sense that it is ever more taken up with the kind of machinery that it itself is' (Connor 2016a: 18).

But, as Purdon observes, while 'writing is technological, through and through, *technology is also written, through and through*'. Technography therefore 'seeks to bring us back to that symbiosis of the technical and the symbolic that is written into the idea of technology from the beginning'. It is 'less a method than a resistance to settling at one or other end of the object-discourse axis' (Purdon 2016). Sean Pryor and David Trotter accordingly define technography as 'a reflection upon the varying degrees to which all technologies have in some fashion been written into being'. But reversibility is again apparent in technography, in that it is attuned not only to the ways in which technologies are written, but also to the ways in which they in themselves constitute styles, or kinds of writing. As such, technographies 'attend equally to the rhetoric sedimented in machines, to machines behaving rhetorically, to rhetoric that behaves mechanically, and to rhetoric behaving in pointed opposition to mechanism' (Pryor and Trotter 2016: 16). As James Purdon concludes, '[t]echnography does not presuppose where, if anywhere, text begins, and the machine stops' (Purdon 2016). I consider in this book a number of different ways in which every technology may be thought of as textual, and all machines as kinds of media. I have distinguished one of these ways:

> If all writing is a kind of machinery, why might it be plausible to see every machine as a kind of writing? Because every mechanical or technical action can be seen as a procedure as well as a mere proceeding, where a procedure means a replicable operation. So a technical procedure is the styling of a process and, as such, the declarative performance of that process as iterable procedure. Every machine declares of what it does: 'this is the way this action may be performed'. A technic, or technical procedure, is a writing in that it tracks its own technique, following in its own tracks. (Connor 2016a: 18-19)

In the particular kind of writing about technology or mechanical rhetoric with which this book is concerned, namely writing about imaginary machines, or machines that do not in fact exist, these convergences and inversions are unignorable. In such cases, the writing of the machine must always be more than merely about it, since the writing of the machine is all the machinery that there in fact is. But, as I have thought about various kinds of imaginary machines, teleportation machines, perpetual motion machines, cyberoneiric machines, influencing machines, time machines, and other devices as yet inexistent, but still obstinately whirring away in imagination, I have become more than ever convinced that there is a continuing work of imagining involved in all machines and indeed, that all machines are in fact imaginary machines. By this, of course, I do not mean that no machines in fact exist. I mean that the acts and arts of imagining, and the spectrum of comportments and affective investments they convoke, are essential to the ways in which machines are 'existed', in Sartre's transitive usage – that is, made to exist, brought to and kept in existence, made liveable, and even, often, loveable.

All machines are contingently imaginary because all machines need to be imagined. Machines must be imagined not just as part of the process of their invention and design, but as part of their habituation and use. This is true because machines have to be designed, engineered, and maintained, but also because we need to learn their use, and to adapt and engineer ourselves, in order to make use of them. Languages like French and German perhaps make it easier to discern and represent this process than English. German offers in the word *Technik* a coalescence of two ideas which are separated in English into technology and technique, where the latter is the way in which one uses the former. Ironically enough, in French usage, the word *technique* allows for both the associations of *Technik*, so it is broader than English *technique*, where the word means the skilful manner of doing something rather than the external means of doing it.

The word *technics* has sometimes been proposed to bring together the external forms and the internal experiences of technology as *Technik* and French *technique* do. Perhaps there might also be some utility in the word *technesis*, which might be thought of as a technical term for mechanical imagining. I would want to distinguish *technesis* from *technoesis*, which names the ways in which technology blends with consciousness in what Roy Ascott has called 'technoetic culture' (Ascott 2003: 353). Technesis would even in one sense be the reverse of this, concerning as it does, not the way in which technology makes mind knowable, but the way in which technology itself is made known and knowable, as a kind of lived knowledge. The word is formed on the model of *allegoresis*, which means something like the application of allegorical modes of seeing. I want the word to mean a little more than Mark Hansen does, when he defines technesis as 'the putting-into-discourse of technology' (Hansen 2000: 20), which he regards as a reduction of technology to text, which undesirably 'purges technology of its materiality' (88). Rather, we might see technesis as the cultural reinternalisation of the externalisation of our capacities we bring about in technology. It is not so much the technological mediation of experience as the mediation of technological experience.

Ingenuity

Imaginary machines are not quite the same as imagined machines, though of course they can overlap. An imagined machine is an imaged machine, the process by means of which a machine that does not yet exist is brought into a condition of discernibility, either in the eye of the mind or through some form of external description or visualisation. Such work of concrete imagining is an important, even a necessary part of the development of any machine. Indeed, that the making of a machine begins when conception moves into contrivance, which must always involve visualisation, as the first stage of making actual. There are cogs in cogitation.

But an imaginary machine is something more than this. An imaginary machine is a fictive machine, not just one that has been postulated. Imaginary numbers, numbers expressed as the square root of a negative quantity, are different from imagined numbers, since they will never become actual as numbers. An imagined country might well turn out to be in existence somewhere; an imaginary country never could. The word 'imagined' is orientated towards the object to be imagined: the word 'imaginary' is orientated towards the work of imagining.

The two are delicately contrasted at either end of a stanza of Shakespeare's *The Rape of Lucrece*, which evokes the way in which a painting of the siege of Troy provides synecdochic hints to be completed by the viewer's imagination:

> For much imaginary work was there:
> Conceit deceitful, so compact, so kind,
> That for Achilles' image stood his spear
> Griped in an armed hand; himself, behind,
> Was left unseen, save to the eye of mind:
> A hand, a foot, a face, a leg, a head,
> Stood for the whole to be imagined.
> (Shakespeare 2007: 224)

Shakespeare's 'imaginary work' does not mean that the work concerned is not real, but rather that it is the work of and for the imagination. Indeed, Shakespeare's own evocation of this painting is imaginary as well as imagined, if only because it is given in words that make us imagine it rather than in pigment that would let us see it (that is, imagine we do). 'Imaginary work' hints not just at the work required to imagine something, but also at the possibility that that the work of imagining is a work that is itself *under imagination*, as we might say, on the analogy of the things, like roads or websites, that are described as 'under construction'. 'Imaginary work' makes us work at imagining the work of imagining. The imagining work gets to be imagined (needs to be), along with what it imagines.

When the word 'imaginary' is used as a noun, to mean a style or distinctive set of imaginative procedures, in phrases like the 'Gothic imaginary', the 'spatial imaginary', the 'oral imaginary' (LaBelle 2014), the 'transplant imaginary' (Sharp 2014), and even the 'technological imaginary' (Punt 2000), it often partakes in this duality, since it will usually mean at once what is imagined and the way in which it is imagined, so signifying both the issue and the operation of imagining. When Michael Punt evokes the 'technological imaginary' of early cinema, for example, he means the set of social and economic conditions that make possible, in the sense of making it possible to imagine, the specific technological inventions and developments of cinema:

> innovations are not the outcome of an internal property of technology, but are the consequences of contingent responses to the conflicting imperatives of both individuals and groups relative to the material state of technical possibilities [...] [T]hese forces must be satisfactorily stabilised and integrated before a technology (whether at the level of the individual artefact or the general system) can arrive at an accepted cultural meaning which, in turn, is one of the preconditions for its social uses and economic success. (10)

This sounds very much like the account of some prior, priming mechanism which is required for the secondary mechanism that is the technology of film to be set going. The mechanism must be socially imagined, through a process that is imagined as itself a kind of mechanism.

So imaginary work is work that imagines itself in operation, indeed, imagines itself as an operation, as though it were itself a kind of imaginary machine, or machine of imagining. Perhaps this reflexivity is always at work when we imagine a machine, since in doing so we might always also be drawn to imagine the work of imagining itself. And it so happens that to imagine this work, we will commonly draw on or dream up machines, devices, instruments, techniques and technologies of all kinds, imagining imagining, for example, as a process of drawing, shaping, projecting, breeding, brewing, weaving, etc.

The mixing together of machinery and imagination is paralleled in the family of words to which engine and engineering belong. Latin *ingenium* signifies an inborn genius, from *in* + *genere*, to engender or beget. By the middle of the sixteenth century, the word *ingenious* starts to be used not just to mean possessing wit, intelligence or good sense, but also to signify the capacity of invention, or of giving rise to other things. At around the same time, it also starts to be applied to the contrivances themselves: so both an inventor and his invention might be called *ingenious*. The same movement of meaning is found in the word *engine*, which, from the 14th century, meant artfulness and ingenuity, as in Gower's *Confessio Amantis*: 'Tho wommen were of great engyn' (Gower 1901: II.367). Chaucer's Parson tells us that 'Goodes of nature of the soule ben good wit, sharpe vnderstondynge, subtil engyn, vertu naturel, good memorie' (Chaucer 2008: 302), while St Cecilia, in the *Second Nun's Tale*, assures her brother-in-law Tiburce that God has three persons, just as 'a man hath sapiences three –/Memorie, engyn, and intellect also' (266). So, in the late medieval period, the engine is both the contriving, the inborn quality of mind that gives rise to some contrivance, and the externalised contrivance itself. The *Medulla Grammatice*, a collection of early fifteenth-century Latin-English glosses in a manuscript in Stoneyhurst College, explains the Latin word 'machinosus' as 'ful engines' (*Medulla Grammatice* n.d. 39 a/b).

If the word 'engine' has subsequently moved from the faculty to its material products, so the contemporary term *search engine* seems to have gone in the opposite direction, from material form to abstract operation. Nobody really has in mind any kind of material arrangement or contrivance when they refer to a 'search engine', a term that is witnessed as early as 1984, but by around the middle of the 1990s was being used almost entirely for accessing internet information. The term 'engine' in this phrase has an oddly archaic feel, but may derive from the 'Automatic Computing Engine', a calculating machine designed by Alan Turing for the National Physical Laboratory in 1945 that has a claim to be the first computer. The term 'Automatic Computing Engine' was devised by John Womersley, superintendent of the Mathematics Division of the NPL (Copeland 2012a: 37, 83), and it is assumed that this was in conscious homage to Charles Babbage's two calculating machines, the Difference Engine and the Analytical Engine, only the first of which was actually constructed (Campbell-Kelly 2005: 156). Turing is said to have despised Womersley, but agreed that the name was well-chosen (Numerico 2012: 177). The link with Babbage is suggested by the fact that one later researcher mistakenly referred to it as the 'Analytical Computing Engine' (Vowels 2012: 226). Turing's paper of 1945 refers to an 'Electronic Calculating Machine', nowhere alluding to Babbage, or to any sort of engine (Turing 1945).

The lingering kinship of engines and ingenuity seems to have encouraged the formation of the portmanteau-word 'imagineer' in 1940s America, a usage which was appropriated by the Disney company, with reference in particular to theme-park entertainments. Disney's idea of Imagineering is related closely to the actualising of the magic of Disney movies in theme parks, first conceived by Disney in 1951. The aim was that 'visitors who stepped into this new park should feel as though they stepped into a movie' ('The Imagineers' 1996: 11). Disney set up a company called WED Enterprises (the initials standing for Walter Elias Disney) in 1952 to design and build Disneyland, the name being changed after the recruitment of Michael Eisner and Frank Wells in 1984 to Walt Disney Imagineering. The term 'imagineering' may have been devised by the Aluminium Company of America, who placed an advertisement in *Time Magazine* for 16 February 1942 that read

> It takes a very special word to describe making aluminium cheap, making it versatile, finding totally new places to use it and then helping people use it where they should. In war times it takes a very special word indeed to describe, also, the ingenuity and daring that can make, almost overnight, three and four and five times as much aluminium as was ever made before, and make it cheaper than ever.

> IMAGINEERING is the word. What aluminium did for civilians, what aluminium is doing for our armed forces, what aluminium will do in the future, all come out of that one word.
> ('Alcoa' 1942: 59)

The advertisement concludes that 'Imagineering is letting your imagination soar, and then engineering it down to earth', implying that the engineering comes after the imagining, but there is room in the term for the idea that it is imagination itself and in the first place that is being engineered. Disney filed for copyright in 1967, claiming first use of the term in 1962 (Kerzner 2014). Despite Disney's appropriation of it, 'imagineering' has found employment elsewhere, for example in Serge King's *Imagineering for Health*, a book explaining the process of 'creating a state of health in yourself, using your own spiritual, mental, and emotional resources. "Imagineering" is the term I use for doing that, because it implies the process of building something with your mind' (King 1981: vii). The book offers training in the use of various tools, including the 'Tool of Imagination', the 'Tool of Motivation', and the 'Tool of Concentration' in order to prime and activate forms of self-healing. Imaginary engines and engineering will frequently be accompanied, in what follows, by the engineering of imagination.

Mechanical Feeling

Machines are imaginary, not just because we have no choice but to imagine their operation and uses, but also because we invest so strongly in them. When we imagine machines, we are doing more than just representing them to ourselves, for imagining is representation coloured and contoured by feeling. What we feel about machines is thrown into sharp relief by the fact that machines themselves seemingly by definition do not feel. This means that we are in a relation of reciprocal surrogacy with machines. Machines do things on our behalf, without needing to have any feelings about it. If machines do not feel, then we do, and our feelings may well be feelings about that disjuncture. Machines do the doing we cannot, we feel the feeling they cannot. Machines give us ways of feeling about ourselves, just because, we are almost sure, they do not themselves have ways of feeling about themselves. We feel for them, and perhaps also through them. What we feel for and about and through them confirms us by recoil in our way of existing ourselves as non-mechanical.

If all technographies, the ways we write, represent, and interpret machines, are suffused with affect in this way, they can all therefore be thought of as psychotechnographies. The feelings we have about machines – as it may be, rapture, fascination, awe, frustration, boredom, rage, terror, even at times a kind of tenderness or pity – are not only intense but also complex, and for this reason are also the occasion for unremitting work. We do not merely 'have' feelings about the workings of machines, we also undertake a kind of feeling-work – on the model of Freud's dream-work, (Freud 1953-74: 5.640) and joke-work – or affect-engineering. Feelings are not just what connect us to machines, or give them a sort of adjectival aura or caption; machines are also what connect us to our feelings and the way in which we work on them and imagine them at work in and on us. We might borrow here the name of a particular nineteenth-century apparatus, the polygraph, so-called because it detected and traced simultaneously many different kinds of physiological data – heartbeat, perspiration, respiration, etc. Perhaps machines, actual and quasi-actual, may all be thought of as polygraphs, ways of mediating and monitoring our states of feeling. When we speak of feelings, as opposed to thinking hard about them, we tend to assume that, unlike thoughts, we simply undergo, or even simply are them, rather than having them as representations. But the machinery of representation is in fact necessary to the ways in which our feelings exist, or are existed.

How, then, do we feel about machines? This question could mean two things: what do we feel about machines; or by what means do we feel what we do about machines? One of my favourite limericks, attributed to Maurice E. Hare, might help us with both:

> There was a young man who said 'Damn!'
> It is borne in upon me I am
> But a creature that moves
> In predestinate grooves:
> In fact, not a bus, but a tram.'

Why is this funny? Partly, I suppose, because of the dummy subject introduced by the construction 'It is borne in upon me', along with the suggestion that the 'I' is in fact itself a machine of conveyance by which things are 'borne'. It is also because the philosophical young man cares about his predicament, or seems to, and we don't, or don't need to. There are some alternative versions that don't seem to me to be quite as funny, though all of them have the first lines essentially as I have given them. We can imagine these first lines instructively disimproved, for example into

> There was a young man eating jam
> As he glumly reflected 'I am...

The jam (slightly more comic than 'ham', I'd say, but there's not much in it) has its own bathetic silliness, for the idea of the consolation it might be offering, especially with its blunt sound amplified in 'glumly', that I find quite pleasing. But it doesn't seem to me to be as funny as the line it replaces because it loses the young man's flare of annoyed offence when he realises his automatism: he's not just cast down, he's positively put out about it. But even this would not be as funny if it did not occur in quite the place it does in the rhyme-scheme

'There was a young man eating jam / Who said 'Damn and blast it! I am...'' This is OK in its way, but still groggier than the canonical version because it lacks the wrench between the sudden intemperate spurt of irritation, which threatens to pull the whole poem off the rails, and the obedient way it immediately settles back into the diddly-dum dactyls of 'But a creature that moves / In predestinate grooves'. Oddly enough, the protest is more trivialised than dignified by the word 'damn', a word we imagine this young man might also use if he had merely left his umbrella on the tram. And yet I also think this is part of the reason that it flips the comic switch more neatly than, say: 'There was a young man who said 'Fuck! / For I see that I am, curse my luck,/But a creature that moves...'' The tension between the predictable rhyme scheme and its unpredictable content, the limerick's standard partytrick, is made more enjoyable by the fact that the content is concerned with predictability. So the effort of the content to break free from the constraint of the container in fact only locks container and content together more tightly. The verse tells us that even the young man's protest at his realisation, a protest that we might be tempted to read as a Luciferian or Blade-Runner assertion of his existential freedom in the face of his predestinate condition (predestined I may bloody well be, but I reserve the right to resent it), is itself part of the programme. So the kinetic energy of the poetic mechanism is what imparts the impetus that might seem to permit escape from it.

Henri Bergson has taught us to see something mechanical at work in all laughter, though he is not sure whether the laughter is part of the machine or not (Connor 2008). 'We laugh', declares Bergson, 'every time a person gives us the impression of being a thing' (Bergson 1911: 58). Yet we do so 'involuntarily' and 'it is really a kind of automatism that makes us laugh' (12, 16). When we imagine machines, and, especially if in the process we imagine our relation to machines, or reflect on our own potentially mechanical natures (and perhaps we must always to some degree do both), we will be doing some affective as well as some representational work. When we think about machines, we are often also forming feelings about them, feeling out how we should feel about them. And precisely because there is a sort of labour in feeling, reflecting on machines provides a kind of laboratory of feeling. In that reflection, the question of how or whether our feeling-work is itself mechanical (and then the further question of how we are to feel about *that*) may always be at work.

The Absolute Machine

Our strong investment in machines has two dimensions. On the one hand, we want to explicate their workings, to work out how they work, thereby demonstrating our priority and superiority over them. On the other hand, we want those workings to be absolute and autonomous of us; we want them to be implicit or hidden, we want them, as a proof of this independence from us, to work by themselves, because, unlike us, they can't get tired, or distracted or discouraged. Our dream is that they will work like a dream, since we can so rarely work our dreams (and, as we will see later, most machines in dreams do not in fact work very well). The last thing we do following the repair or maintenance operation is to replace the cover, whether of the car, television, or computer. We image machines the better to imagine them working without our needing to hold them in view. Machines are imaginary because we want to imagine them as independent of us. We want to work them, while also wanting them to work on their own. Machines depend on us for the independence from us on which we depend.

As such, all machines are liable to suggest ways of imagining our own bodily and psychological mechanisms, which we similarly possess and control but do not run, nor wish to have to. All machines have the capacity to form on our behalf a kind of self-image, of a machinery we cannot fully conceive or encompass, but is us in our ideally self-maintaining form. Machines suggest ways of imagining the unimageable machinery we take ourselves to be, the workings of lungs, livers, immune systems, hormone cycles, and neural networks, but also dreams, desires, and aspirations. Most importantly, machines allow us to maintain the ambivalence that enables us to be sure that we both are and are not machines, that we know how to work ourselves, even though our selves work, as we say, by themselves.

So the first machine, and the machine of last imaginary resort, is not in fact any actual machine, but the imaginary machine of the human economy, even if we have to be taught about it from the outside in by the secondary machines we meet and make in the world. This is made explicit in Arnold Bennett's *The Human Machine* of 1908, which begins by evoking the passionate involvement of inventors:

> They are continually interested, nay, enthralled. They have a machine, and they are perfecting it. They get one part right, and then another goes wrong; and they get that right, and then another goes wrong, and so on. When they are quite sure they have reached perfection, forth issues the machine out of the shed – and in five minutes is smashed up, together with a limb or so of the inventors, just because they had been quite sure too soon. Then the whole business starts again. They do not give up – that particular wreck was, of course, due to a mere oversight; the whole business starts again. For they have glimpsed perfection; they have the gleam of perfection in their souls. Thus their lives run away. 'They will never fly!' you remark, cynically. Well, if they don't? Besides, what about Wright? With all your cynicism, have you never envied them their machine and their passionate interest in it? (Bennett 1911: 8-9)

Bennett identifies an envious longing for absorption in a perfect, or rather, endlessly perfectible because imperfect, machine: 'have you not wished – do you not continually wish – for an exhaustless machine, a machine that you could never get to the end of?' (10). The idea of a machine that is 'exhaustless' because it is always going wrong, is a striking insight, and we will meet this idea again when we consider the pleasure given by machines. In fact, Bennett urges his reader, every person does indeed possess such a machine:

> It has never struck you that you do possess a machine! Oh, blind! Oh, dull! It has never struck you that you have at hand a machine wonderful beyond all mechanisms in sheds, intricate, delicately adjustable, of astounding and miraculous possibilities, interminably interesting! That machine is yourself. (10-11)

What, on this view, is a human? A human is a creature who dreams himself up in and as machinery. *Homo mechanicus.*

It may well be said in response to Bennett's view that not all humans are like this, and that, indeed, a broad distinction can be made between bad, cruel, and lifeless people, of whom there are far too many, and who are possessed in some unholy fashion by the idea of machines, to the point of becoming themselves machine-like; and good, harmless, vulnerable, blessed people, of whom there are too few, and who find machines menacing and dull and incomprehensible, thereby confirming their living humanity. The idea that human beings are at risk of losing their humanity to machines and to mechanical conceptions of themselves has, since the rise of industrialism, become a kind of automatic reflex, about which we scarcely need to think, since it thinks itself for us.

Yet even the most devoutly immovable antimechanist will find it hard to maintain this negative view consistently. Humans, even and perhaps especially those who seem to loathe them, like D.H. Lawrence, are fascinated by machines and mechanisms; it is no accident that they are often also thought of as toys and playthings, and vice versa. Nobody with any life in them is capable of simply 'using' a machine. Perhaps this is because all machines seem to suggest some kind of state of exception to the rule that material things may be moved by other things but are not capable of moving themselves, and so are not, in other words, automata, in the early, exact sense of the word, that they move, or move by, themselves. We regard material things that can move themselves as living creatures and are correspondingly intrigued by the anomalies we know as machines, which can apparently move by themselves and yet are not living. The intensity of this sense of anomaly has dimmed somewhat as a result of the fact that the word 'automaton' has actually come to mean something like the opposite of self-moving – thus, an automatic or automatistic response is one that is not in your conscious control, but rather one in which you are moved by some alien agency, like a mesmeriser or an obsessive compulsion. But the fascination and delight of employing and contemplating a machine is that, as we say, it 'works'.

Machines work on their own, and something that works on its own, without meaning to, and without the power not to, seems to be just what we have in mind when we think of a machine. Machines are, as we say, *automatic*, they are literally self-moving, only by an immanent rather than transcendent will. So machines do not in fact move themselves, as the word *automaton* might suggest, since that would require some internal division plus relation between an active and passive aspect, but rather move *by themselves*, where the preposition *by* might be taken to suggest something like 'by means of themselves', or through their own agency. Machines seem to have an undivided as opposed to a divided reflexivity, a reflexivity without relation. A non-machine has an intent which may be separated from its action, it can be something other than what it does, whereas a machine is what it does, without remainder; a machine just is its purpose in action. Thus, even when they involve self-monitoring and feedback mechanisms, machines are insensate, deaf, dumb, and blind, purely objective. They may seem to know what they are doing, but they do not know that they are doing it. They have nothing to do with us, precisely because they can have nothing essentially to do with themselves.

And yet perhaps this cannot be true, as we can appreciate if we reflect on the difference between a tool and a machine. A tool works on the world: it is a mediated way for a human or other agent to perform work upon the world. A machine will very likely work upon the world too, but will also work on itself, through its internal workings. A tool has a relation only to its user and its purpose. A machine has this too, but also in this sense has a relation to itself. A tool is a verb, a machine is a sentence.

And then this is the very reason that machines in fact have everything to do with us. Machines are mobile arrangements that have no idea of themselves. We, by contrast, we unmechanical entities, entities taking themselves to be nonmechanical, are taken up with ourselves, and for that reason are also taken up by the idea of machines, as a way of giving us our idea of ourselves. Machines, which do not and cannot mediate themselves, mediate us to ourselves precisely through their immediacy, through the strange allure of the idea of the immediate they mediate to us.

It is not possible to be neutral about machines, since the field of feeling-force that surrounds the idea of machinery in general means that even one's neutrality will always in fact be highly charged. Machines themselves are unfeeling, of course: you cannot hurt them, and they cannot care about the ways in which they hurt us, which seems to be precisely the reason why we spend so much time feeling such complicated, watchful things about them. Watchful, because it is important to prevent one's feelings about machines from being contaminated by them; one must avoid at all costs one's feelings becoming mechanical, for this would of course make them no longer feelings at all. Yet, if there is a logic to our feelings, how are we to prevent our feelings about machines becoming feeling-machines?

In a certain sense all living things can be regarded as putative machines, the mechanisms of which are invisible or difficult to imagine. But more importantly the fascination of the machine is that it rhymes so closely with our experience of ourselves, and in particular the fact that I can and must do what I do, think, wish, feel, move, without knowing how I do it. How do I make my fingers move over the keyboard and press the right letters to form these words? I just will it and, pretty much all the time, *it works*. But I have no idea *how* I will it, by which I mean I literally can conceive no image of the process I am setting in train when I will an action. The most mysterious of these actions are mental operations. I know how to think, I can do it actively, directedly, and sustainedly, sometimes for minutes at a time, well, let's say very nearly a minute. But how do I know how to do it, and how exactly do I do it when I do? I am less the ghost in the machine than the *persona ex machina*, the ghostly operator-effect of an unimaginable machine that knows much better than I do how it works, precisely through not needing this knowledge in order to work. Actually, it is not quite true that I have no idea how my (sometimes unwilling) willing-machine might actually work, precisely because I have seen so many machines in operation. Whenever I try to understand concretely and empirically the way in which I, as a self-electing exception to the law of mechanism, exceed the condition of a mere machine, I need to revert, if only dimly, to imaginary mechanisms.

On the face of it, a machine is a closed and finite thing, a proceeding that has a definitive beginning and end, because we may define a machine as something that is absolutely and without remainder what it does. A machine is nothing more than its machinery, or mechanical action. This quality of identity between being and action may be the best definition that can ever be given for a machine.

And yet such self-identity is, however, always ideal or imaginary. For there is something in every machine that falls short of or exceeds this autogenic self-identity. Every machine, even one as simple as a screw or mill, is in fact only one of many possible instantiations of its principle of operation. Every machine can be improved, though the means, or, as it were, secondary mechanism of that improvement is human ingenuity. All machines are in a sense both exemplary and exemplificatory. All machines are more or less imperfect surrogates or figurings of the absolute machine, or an absolute idea of machinery. Just as Jürgen Habermas imagines all utterances as promises or proleptic fragments of an 'ideal speech situation' (Habermas 1984: 25), so every instance of the mechanical seems to be an anticipation of an ideal mechanical apparatus that would work without limit or impediment. The absolute machine would transcend time and undo the Second Law – a perpetual motion machine is a machine for overcoming the fundamental imperfection of the machinery of the cosmos, namely that it is subject to time.

The ideal or absolute machine would also be absolutely independent of our use of it, and independent too of any particular kind of embodiment. Perhaps in a sense, all machines are synecdoches or preliminary sketches for the absolute Machine, the machine that inhabits and animates human thought, the Machine that would in fact be identical with thought itself. At that point of comprehensiveness and frictionless self-equation, the Machine might just as well be thought of as God.

An imaginary machine is more than just a way of conceiving or projecting machines that do not yet exist, though my larger aim in reflecting on the importance of imaginary machines is to show that there must be something imaginary in all machines. An imaginary machine may also be a way of imagining other kinds of thing that a machine can do and be. And since machines are the way in which we come to ourselves through going beyond ourselves, imaginary machines are ways of mediating that mediation, a simulated return on the investment we make in ourselves in investing those selves in machines. In this way, as in others that we will meet, all machines may be said to be media, for they allow for a particular kind of mediation to ourselves, we seeming non-machines, of the idea of the machine.

This Machine Is To Him

The history of the ways in which machines are both written *as*, and in themselves a writing *of*, the idea of the machine, is also the historical generation of feelings (of desire and dread, for example, desire for the machine and dread of it, and so the desire of dread which may seem to protect against the dread of that desire). The history of dream machines is therefore also a history of the machinery we use to imagine the kind of dreaming, dreading thing a self might be. Accordingly, my proposition will be that every technography, insofar as it is of any interest to us, must also be a psychotechnography, and every psychology an implied or imaginary technology of self. This is to agree with Bernard Stiegler's principle, explicating the palaeontologist André Leroi-Gourhan, that 'it is the tool, that is *tehknē*, that invents the human, not the human who invents the technical. Or again: the human invents himself in the technical by inventing the tool – by becoming exteriorized techno-logically' (Stiegler 1998: 141). The human comes into being, not, in Stiegler's terms, as the *who* hauling itself clear of the *what* (*Wo Es war soll Ich werden*, in Freud's optimistic, if also much-mauled formulation, Freud 1991: 15.85), but as their 'co-possibility, the movement of their mutual coming-to-be, of their coming into convention. The *who* is nothing without the *what*, and conversely' (141). The psyche comes between machines and writing; machines come between the psyche and its writing. Always this is a matter of kinds of reflexivity, in which machines mediate (come between yet also connect) our self-relation, and we mediate the self-relation of machines. Subjects, machines and systems of mediation are themselves mediated by propagated catachresis, in which one's self-relation is formed through oblique substitutions and 'takings for' – the subject is taken for a kind of writing, which is taken for a kind of machine, which is taken for a kind of subject through being taken for a kind of writing.

There is perhaps no need to posit, as Tania Espinoza does in considering the mediation of Winnicott's 'transitional object' in the formation of the self, a 'technical unconscious' (Espinoza 2013: 163); though one might certainly point to a kind of delegation of awareness in the employment of every technical device. To know, or expect, that a device will perform the action it is expected to is to surrender or put aside some portion of one's conscious supervision, to put oneself in a limited form of abeyance, by incorporating some automatism, some thing that does not need to know what it is doing. But this is not any kind of second self, a psychic reservoir brimming with unknowns, or 'constantly threatening to dissolve the fragile "internal milieus" that still constitute subjects' singularity' (163) – it is simply the unconsciousness that is intermittently at work in every relation to a techne or technique, insofar as it does not need to be known or made consciously present. Humans are highly productive of such ideal or limitless machineries, which are at work in all logical systems, whether metaphysical or mathematical. Such machines would emanate from and encompass us, even as they exceed us. Indeed, the self-exceeding machine would be a model of the human exceeding itself in its machines and in its idea of machinery, its proleptic *technesis* of the techne.

Busy as his work and imagination are with machinations, Shakespeare uses the word 'machine' just once. Polonius is reading out to Gertrude a letter of Hamlet's to Ophelia. In his letter, Hamlet breaks off from the bit of doggerel he has been composing for Ophelia, protesting that he is *'ill at these numbers, I have not art to reckon my groans, but that I love thee best, most best, believe it'* (Shakespeare 2005: 246). The salutation with which he ends the letter repeats this message of devotion: *'Thine evermore, most dear lady, whilst this machine is to him. Hamlet'* (246). Hamlet's words swivel between two meanings. 'While this machine is to him' is usually taken to mean, 'while this physical frame is his'. The Arden edition from which I quote provides a full stop before the name with which Hamlet signs himself. But, even with this full stop, it is still possible to absorb the name into the sense of what immediately precedes it, thus: 'while this physical frame is to him "Hamlet"'; 'while this machine is what is understood by "Hamlet"'; 'while this body is the way in which Hamlet understands himself to be "Hamlet"'. Of course, alas, one must go further. For 'this machine' may refer, not only to the body that is performing the action which engrosses or composes it at this point, namely, writing a letter; but also to the action of writing itself, an action that expresses to Hamlet the idea of a machinery producing the idea of 'Hamlet'.

We pass through the relay of the machine, the idea that the machine incarnates, in order to round upon ourselves. By helping us be sure of what we are not, machines help us *come to*, as we say ('or rather *from*', as Beckett puts it). Because we are such incessant proxy readers of machines that cannot read themselves, and we write and read ourselves off from them by a logic of exception, we appear to ourselves *ex machina*, rendering ourselves machine-readable.

In the three volumes of his *Technics and Time* Bernard Stiegler has emphasised the importance of temporality in this act of mediation. Technologies always represent the possibility of anticipation, or a purposive orientation to a future time: 'anticipation *itself supposes the technical object*, and no more precedes it than does form matter' (Stiegler 1998: 81). Technologies are the embodiments of intentions, or stretchings-toward (the word 'intention' contains etymologically the suggestion of an instrument, Latin *tendere* being used to refer to the bending of a bow). They are also, of course, inasmuch as they are externalisations of human capacities and intentions, capable of surviving to bear witness to them. In mediating between future and past times in this way, technologies make the human experience of temporality as a kind of tension possible. Hamlet's machine can always point the way to Hamlet's machine – programming a work like Heiner Müller's 1979 play *Die Hamletmaschine*, or at least making it possible for Müller's play to seem like a machine that is to him Hamlet.

It is this reading and writing of ourselves via the reading and writing of machines, including the machines of reading and writing themselves, that my investigation of imaginary machines is intended to allow. In imagining machines, we are able to assure ourselves we exceed or fall short of the condition of mere machinery, even as imaginary machines may nevertheless seem to edge us towards an ever more precise imaging of ourselves. Indeed, the imagining of ideal machines may be precisely the mechanism – a mechanism which is itself at once imaginative and imaginary – we need to give rise to ourselves. Imaginary machines are always formed by a machinery of imagination that they come to resemble.

The most potent imaginary machine of all, and perhaps the matrix of all other imaginary machines, is the Not-Machine, or machine-which-is-not-one, *ce machine qui n'en est une*, or, put more simply, Life. Life is the not-machine, which sheathes together two meanings of the Not-Machine: that which is not a machine, that which must be defined as beyond machinery, or of some different order altogether, and the machine that 'nots', that says no to the possibility that it is no more than a machine, a machinery for showing the limits and transcendence of the merely mechanical. The not-machine conjoins the negation of the machine and the machinery of negation. The not-machine is paradoxical and a machine for generating paradox, for it is at once that which refuses the idea that there are, and can only ever have been, machines, and that which confirms it. If, as Bernard Stiegler writes, 'technics is the pursuit of life by means other than life' (Stiegler 1998: 17) the not-machine is the bridge between life and the 'other than life'. The not-machine is the machine that is programmed to say 'I am not a machine'. When the dictatorial director Julian Marsh in *42nd Street* is roared at by his doctor 'Good

God man, you're not a machine', he means that Marsh should remember he is not possessed of infinite reserves and cannot push his body and nerves beyond what flesh will stand: in other words, he should remember that he is not in fact an imaginary machine, or the kind of machine he imagines himself to be. He must not forget that he is in fact subject to physical limits: that he is, in fact, a machine, rather than the limitless 'Not-Machine' he takes himself to be.

The history of technology tends toward and bends round into a kind of psychohistory, or perhaps rather a psychography of machines. Just as technography moves us from an emphasis on how we know machines (the early sense of the word 'technology') to how we write them (verbally, visually, performatively), so psychography might be taken up with the ways in which selves and selfhood are graphically mediated, with techniques and technologies being one of the principal ways in which this occurs. So technology – the idea of technology, the feelings engineered in the idea of technology – is the self's manner of writing, or making itself known to itself.

There is no chapter in this book on the android. There might well have been, but this might have been to create an artificial separation between machines that may be regarded as andromorphic, in that they figure or image human being, and machines that are not, in that they merely exercise certain functions. But no machine can figure the human in itself, since this figuring is always projective, a kind of postulation, which will always be too early or too late in respect of the dreamed co-implication of the *who* and the *what*, *psyche* and *techne*. If all machines are in fact androids, since all in fact serve to figure the ongoing self-figuration of humans as and through machinery, there can be no coming to rest in a final convergence of human and machine, however charged with fantasy this may be, and, in fact, precisely because it is so charged with fantasy, that is, charged with the duty of being and remaining unrealised.

So perhaps the varieties of machine considered in the chapters that follow may be thought of as collectively constituting a kind of polyandrous homunculus, in which different human, or prospectively human faculties, of moving ('Transports'), dreaming ('Dream Machines'), desiring ('Pleasure Machines'), healing ('Medical Machines'), influencing ('Radiation Machines'), disappearing ('Invisibility Machines') and persisting in perpetuity ('Perpetual Motion', are figured. I begin with the machineries, actual and imaginary, that mobilise the human body.

2

Transports

Transmission

Machines transmit force. A windmill captures and transmits the force of the wind, just as a water-mill captures the force of flowing water and transmits it to a wheel that grinds against a stone. An Archimedean screw lifts water by converting rotary force into upwards force. Uicker et. al. define a machine as 'an arrangement of parts for doing work, a device for applying power or changing its direction' (Uicker et. al. 2003: 6). They quote the definition of a machine offered by Franz Reuleaux in his *Kinematics of Machinery* of 1875: 'A machine is a combination of resistant bodies so arranged that by their means the mechanical forces of nature can be compelled to do work accompanied by certain determinate motions' (Reuleaux 1876: 35). There are two cooperating ways in which force is applied in a machine. Force is applied by the machine to channel or redirect the dynamic powers at work in nature in order to apply this redirected force to a particular task. So the sails of a windmill convert the energy of a current of air into a rotary motion. The first application *forces transmission*. This movement is then itself converted, in the case of a windmill or watermill through a system of cogs that changes the plane of the rotation, into a means for grinding one stone against another, in order to grind grain, for example. This second application *transmits force*.

A machine depends both upon the clear distinction of the components through and by means of which force is directed or applied – wheel, crankshaft, piston – and the combination of those components to create motion which moves across the machine. A certain kind of geometrical mathematics is needed to conceive and build any machine, but this machinery is also itself suggestive and productive of a kind of mathematics. Machines are built of mathematics, but machines also build mathematics. Machines often require models, but are themselves modellings. So it might be possible to say that the experience of machines was needed to form and convey the very idea of force, as a measurable and comparable quantity. Reuleaux, who was the first great systematiser of the science of kinematics, even seems to suggest that machines themselves may be regarded as components in an abstract work of thought which, analysed itself as a kind of logical machinery, allows for the production of new machines. All machines, however elementary, 'have been thought out by human brains, – now and then by brains of special capacity, and then praised as God-sent gifts, but in all cases *thought out, produced by a mental process which has contained more or less well-defined gradations*' (20). Reuleaux saw it as the task of what he called kinematics to externalise the '*mental process*' involved in the invention of any machine (20). For this he developed an abstract topological system, which analysed the elements in a machine into sequences and networks of kinematic pairs, each of which both constrained and transmitted force. He described this as a 'phoronomy' or a geometrical study of motion (56). Invention is the movement supplied by thought. By generalising the theory of mechanical movements, Reuleaux aimed to generate a machinery of thought – of which, of course, a new kind of kinematic notation was a central element. Understanding the transmission of force therefore enjoined a technographic operation, in that it was a problem of representation and translation of action into sign.

Reuleaux built on the work of Charles Babbage who found it necessary to develop a symbolic notation to understand the mechanical operations undertaken by his calculating machine. Babbage described in 1826 the difficulties that led to his method of notation:

> The difficulty of retaining in the mind all the contemporaneous and successive movements of a complicated machine, and the still greater difficulty of properly timing movements which had already been provided for, induced me to seek for some method by which I might at a glance of the eye select any particular part, and find at any given time its state of motion or rest, its relation to the motions of any other part of the machine, and if necessary trace back the sources of its movement through all its successive stages to the original moving power. I soon felt that the forms of ordinary language were far too diffuse to admit of any expectation of removing the difficulty, and being convinced from experience of the vast power which analysis derives from the great condensation of meaning in the language it employs, I was not long in deciding that the most favourable path to pursue was to have recourse to the language of signs. (Babbage 1826: 250-1)

Babbage had been much taken up as an undergraduate in Cambridge with the problems and possibilities of different modes of notation, and made strenuous efforts to introduce Leibniz's fluxion notation for calculus, believing it to be superior to the Newtonian system adhered to in Cambridge (Babbage 1864: 38). In a letter of June 1852 to the Prime Minister, Edward Smith-Stanley, he made it clear how vital this 'system of Mechanical Notation, by means of which the drawings, the times of action, and the trains for the transmission of force, are expressed in a language at once simple and concise' was to his work on the Difference Engine which he was offering to the government, and asking it to pay for: 'Without the aid of this language I could not have invented the Analytical Engine; nor do I believe that any machinery of equal complexity could ever be contrived without the assistance of that or some other equivalent language. The Difference Engine No. 2 [...] is entirely described by its aid' (104).

Reuleaux went beyond Babbage's system of mechanical notation to develop a system that, given any particular form of machinery, allowed 'for the realization of the abstract form of the mechanism, for the perception of its essential nature under its material disguise' (Reuleaux 1876: 259). Translating machines back into the mental processes of which they were the actualisations would, he believed, allow for the production of new machines:

> If the processes of thought by which the existing mechanisms have been built up are known, it must be possible to continue the use of these processes for the same purpose; they must furnish the means for arriving at new mechanisms, must, that is to say, take up the position hitherto assigned to invention [...] Invention, in those cases especially where it succeeds, is *Thought*; if we then have the means of systematizing the latter, so far as our subject goes, we shall have prepared the way for the former. (20)

Here the operations of machinery are abstracted into general mechanical principles, but in order that a kind of thinking can be formalised as a sort of machinery, allowing in turn for further inventions. So the mechanisation of invention potentiates the invention of mechanism. Actualising the work of the mechanical imagination, turning a hard machine into a 'soft' or imaginary machine, makes it possible to generate new forms of 'hard' mechanism, just as formalising the rules of a language allows one to imagine new language forms.

Reuleaux's work can be thought of as an example of what Peter Sloterdijk has called the work of 'explicitation' (Sloterdijk 2004: 87), intended 'not so much to add to the positive knowledge of the mechanician as to increase his understanding of what he already knows' (Reuleaux 1876: 23). Transmission not only governs the workings of most machines, it is also the central principle of the explicitation of machines. Machines transmit force, but the means and method of that transmission can themselves be made transmissible through symbolic formalisation. The reason that narrative has often been thought to be a kind of mechanism, or to be describable in terms of the movement between specifiable parts, may be that machines themselves seem to have a kind of narrative action, involving components that act and are acted upon in specific and repeatable sequences.

In this chapter, I focus on the movement of force that is a feature of all machines, by paying particular attention to vehicular machines whose function is to provide movement, by transporting goods or persons from place to place. Imaginary vehicles allow the idea of transport itself to be transported, to some other place or modality.

Vehicular

The word *vehicle* seems not to have been in use in English before 1615, when it appears in Helkiah Crooke's *Microcosmographia*, which refers to a vein as 'the vehicle or conueigher of blood' (Crooke 1615: 80). In seventeenth-century English, the word *vehicle*, now used almost exclusively for transport machines, was applied to many different forms of medium or vector (from the same Latin root *vehere*, to carry). The two principal areas of application of the word *vehicle* were in theology and in medical chemistry – it is a common term in the work of Robert Boyle and was also used in medical texts to refer to forms of solvent, sweetening, or emollient fluid in which to convey bitter or unpalatable medicines. It was sometimes assumed that vehicle and conveyance are related terms, but this is an etymological error; *convey* derives from *convoier*, to go along or share a path with (as in the word *convoy*), but the Renaissance spelling *conveigh*, on the model of *inveigh*, created the mistaken impression that, like *vehicle*, it derives from *vehere*. (*Purveying*, which might seem kith and kin to these words, has another derivation still, from *pourvoir*, to provide for.)

I.A. Richards's suggestion that a metaphor can be divided into its tenor, or general import, and its vehicle, or its form, such that '[t]he tenor, as I am calling it [...] [is] the underlying idea or principal subject which the vehicle or figure means' (Richards 1965: 97), has a long history, since the word *vehicle* was often used in the sense of a material embodiment of some more abstract principle: we read, for example, in a seventeenth-century study of government that 'Power is a sort of Volatile Spirit which cannot subsist without a proper Vehicle to give in Body' (A.B. 1689: 17).

This can make the word *vehicle* itself able to be a vehicle for mediating actual and symbolic conveyance, one on which Jeremy Collier plays in his mockery of the poetic speech in Congreve's plays:

> This *Poet's* way of understanding *English*, puts me in mind of a late Misfortune which happen'd to a Country Apothecary. The Dr. had prescrib'd a Lady Physick to be taken in something Liquid, which the Bill according to Custom call'd a *Vehicle*. The Apothecary being at a Stand about the word, applies, as Mr. Congreve might have done, to *Littleton's Dictionary*. And there he finds *Vehiculum* signified several considerable Things. He makes up the *Bill*, and away he goes to the Lady, where upon the Question, how the Physick was to be taken? He answers very innocently; Madam, says he, You may take it in a *Cart*, or a *Waggon*, but not to give your Ladyship too much trouble, I think a *Wheelbarrow* may do; for the word *Vehicle* in the Bill, will carry that sense. In

short, This Direction was comply'd with, and the *Footman* drove the *Wheelbarrow* about the Chamber. (Collier 1699: 56-7)

The idea of a vehicle is therefore likely to be evoked whenever there is reflection upon the ways in which the soul is carried by or joined to the body, for example in this passage of reasoning from Cornelius Agrippa as rendered in John French's 1651 translation:

> mans soul being such, according to the opinion of the *Platonists*, immediately proceeding from God, is joyned by competent means to this grosser body; whence first of all in its descent, it is involved in a Celestiall and aeriall body, which they call the celestiall vehicle of the soul, others the chariot of the soul: [...] thus it is manifest, how the immortal soul, by an immortall body, *viz.* an Etheriall vehicle, is included in a grosse and mortall body, but when by a disease or some mischief, these midle things are dissolved or fail, then the soul it self by these middle things recollecteth it self, and floweth back into the heart which was the first receptacle of the soul: but the spirit of the heart failing, and heat being extinct, it leaveth him, and man dieth, and the soul flieth away with this Celestiall vehicle. (Agrippa 1651: 465-6)

Here, the vehicle is literally a mediation or 'midle thing' allowing for the passage of the spiritual into the corporeal, which is rendered both as a metaphor and as a literal 'chariot of the soul'. The problem is recursive, since some further mediation must be required in order to make it possible for the soul both to be and to be conveyed in its 'celestiall vehicle'.

This draws the idea of the vehicle into the whole choppy history of uncertainty about truth and its potentially dangerous rhetorical mediations. So we should not be surprised to find that the idea of the mediating vehicle can often have demonic associations, for example in a suggestion of 1664 that 'the Divel hateth Musick liberal, and on the contrary is delighted with filthy Musick and illiberal, which he useth as his Vehicle, by which he slideth himself into the minds of men' (Alstead 1664: 4); and similarly in an anonymous pamphlet of 1676 that mocks the belief that 'the same grief or wound, that a Witch receiveth in her aerial, assumed body, or vehicle, may be found and felt too (though the Soul be absent) at the same instant, in her dead Carcass left behind at home by the wall' (*Doctrine of Devils* 1676: 89-90). Sometimes the demonic seems to be embodied in the very facility with which spirits could speak through different kinds of refracting and debasing vehicle. Richard Bovet wrote of the pagan spirit of Apollyon that it spoke its oracles *'per Pudenda Puellae'*, through the genitals of the girls known as *ventriloqui*, and that '[s]ometimes the open Air have resounded with voices, sentences, and noises of this Infernal Daemon, sometimes assuming the Vehicle of one

Shape, sometimes of another' (Bovet 1684: 37). Melancholy was known as *Vehiculum Diaboli* since it gave opportunities for the devil to seize the soul (Jollie 1698: 47). It seems a little strange that Milton should not have found a single occasion to use the word in *Paradise Lost*.

The word *vehicle* was often understood to mean a type or figure. These are the terms in which Scottish Quaker Robert Barclay defended himself against the charge that his use of the phrase 'Vehicle of God' to refer to Christ's body was '[a] wonderful expression, savouring more of a distracted braine, and of an audacious, blasphemous spirit, than of a sober Christian, fearing God (Brown 1678: 233). Barclay replied that

> As for the word *Vehiculum Dei* as having a respect to Christ's Body or Flesh and Blood from heaven, that it is a Scriptur word [...] and that by *Solomon* is mystically understood *Christ*, of whom *Solomon* was a figur, or type, none who are spiritually minded can deny, and consequently that this chariot or vehicle must be mystically and spiritually understood (Barclay 1679: 162)

Often, the defence of the actuality of devils and spirits depended upon the argument that they could act directly on and through matter as their vehicle. In a sceptical discussion of the Salem witch trials, the Bostonian Robert Calef quoted a 'Letter of a Gentleman, endeavouring to prove the received Opinions about Witchcraft' (Calef 1700: 64) which maintained that:

> a Spirit can Actuate, Animate, or inform a certain portion of matter, and be united to it: from whence it is very evident, that the Devil united to a portion of matter (which hereafter I'll call a Vehicle) may fall under the cognizance of our Senses, and be conversant with us in a bodily shape. Where then is the reason or need to run to a Metaphorical, and forced Interpretation, when the words are so plain, and the literal sense implieth no contradiction, nor any greater difficulty than (as has been said) what ariseth from the Union of the Soul and Body, which is most certain. (67)

So, not only do vehicles often function as metaphors, metaphor seems often to call up or call on some imaginary machinery of carriage. Translation between the two modes of translation, the spatio-physical and symbolic-semantic, itself requires and requisitions vehicles.

Many stories or dreams of impossibly swift passage require the imagination of specific forms of carriage. We can often note in such forms of imagined or magical transport a principle of progressive dematerialisation. The machine that is required to increase one's speed, or transport oneself to some otherwise inaccessible destination, is likely to dissolve in the process. A short story by

Kafka, consisting of the following single sentence, enacts with scarcely credible economy the process whereby a form of carriage melts away into pure idea:

> If one were only an Indian, instantly alert, and on a racing horse, leaning against the wind, kept on quivering jerkily over the quivering ground, until one sheds one's spurs, for there needed no spurs, threw away the reins, for there needed no reins, and hardly saw that the land before one was smoothly shorn heath when the horse's neck and head would be already gone. (Kafka 1993: 390)

The speed of the horse's motion is such that the rider, and the sentence that carries the idea of the rider, passes across into pure passage, discarding everything that has made it possible. The single, self-consuming sentence emblematises the many stories of artefacts of transport consumed in the act of supplying movement. It is apparent in Amundsen's successful trek to the South Pole, where a large part of his advantage over Scott's expedition lay in the fact that he fed the weaker dogs who died during the course of the journey to the remaining dogs, in effect creating a kind of feedback mechanism, which allowed his means of transport to consume and convey itself. Writing in 1961, Walter Sullivan compared the Antarctic race between Norway and Britain to the race between the Russians and Americans to reach the moon: 'The moon will be reached by burning up a succession of rocket stages and casting them off. This, in effect, is what the Norwegians did with their dogs, the weaker animals being sacrificed to feed the other animals and the men themselves' (Sullivan 1961: 176).

Not only did it constitute a self-consuming artefact, the Apollo moon landing required what had previously been the nose-cone of the rocket to become a kind of propulsion device, capable of blasting off from the moon and navigating back to earth. It is as though the wheel, or leg, had become pure eye, the eye a principle of absolute propulsion. Space craft are imagined in the same way as flying geese, horses, coaches, cars and rockets, always with the eyes on top of or out in advance of the machinery, in an acknowledgement of the principle that speed transforms matter into appearance. Driving or piloting such a device requires an exercise of bodily imagination which keeps one in charge even though one may be reliant upon mechanical systems – this is particularly true of a large jet plane, where the pilot is required to imagine their body-schema extending out to the edges of a huge machine, in just the same way as a car driver winces at the wing-mirror scraping the garage wall. Such a pilot is like the little boy seated at the window on the top deck of the bus – or, when I was younger, just to the left of the driver on the bottom deck – where one's watching could become one's driving. Why else do plane seats all face the front, unless to symbolise the act of collective and directed volition required to keep the plane aloft and on course? Though there are many who feel uncomfortable about travelling backwards there seems to be less of a necessity for forward-facing seats in trains, cars or coaches. We both wish and do not wish the act of being transported to be voluntary, or to involve mechanism, in confirmation of the principle articulated in Chapter One that machines must depend on us for their independence from us. In this, flight effects the essential ambivalence of every automatic machine, namely that it will drive itself, while being driven. In a machine of transport, the locomotive wheel becomes a steering wheel: driving means steering, rather than producing the movement.

In stories of voyages undertaken in improbable vessels, there is another kind of movement of which account should be taken, from the machinery that sets the story in motion to the machinery of the story itself – or from propulsion to event and adventure. In Francis Godwin's *The Man in the Moone*, which inaugurates the genre of moon-voyage narratives, the flight is effected by a flock of geese, trained by the marooned adventurer Domingo Gonsales. Godwin imagines a machine made up of two components: the trained geese, who really supply almost all the propulsion and much of the direction (they do not have to be induced to fly to the moon, since it turns out they migrate there annually), and what Gonsales calls his 'engine'. This is no more than a system of pulleys employed to persuade the birds to lift his weight all together, as he describes, somewhat less than perspicuously:

> I began to cast in my head how I might doe to joyne a number of them together in bearing of some great burthen: which if I could bring to pass, I might enable a man to fly and be carried in the ayre, to some certaine place safe and without hurt. In this cogitation having much laboured my wits, and made some triall, I found by experience, that if many ware put to the bearing of one great burthen, by reason it was not possible all of them should rise together just in one instant, the first that raised himselfe upon his wings finding himselfe stayed by a weight heavier then hee could move or stirre, would by and by give over, as also would the second, third, and all the rest. I devised (therefore) at last a meanes how each of them might rise carrying but his owne proportion of weight only, and it was thus.
>
> I fastened about every one of my *Gansa's* a little pulley of Corke, and putting a string through it of meetly length, I fastened the one end thereof unto a blocke almost of eight Pound weight, unto the other end of the string I tied a poyse weighing some two Pound, which being done, and causing the signall to be erected, they presently rose all (being 4 in number,) and carried away my blocke unto the place appointed. (Godwin 1638: 24-5)

If all machines are also media, then machines of transport establish this equivalence most visibly. Mary Baine Campbell points to the entanglement of the narrated machinery and the machinery of narration in *The Man in the Moone*, for example in its incessant noting of periods of time and rates of passage, the time taken for something to happen, and the time taken for it to be conveyed to the reader (Campbell 2011: 199). Campbell sees the book as a response to a new world of speed, or rather perhaps world of new speeds, in the late sixteenth century. She concludes that

> [w]here we see superficially, in two dimensions, as it were, oppositions and paradoxes – of logic and fiction, manuscript and print, real travel and visionary dream, poetry and taxonomy, mercantile empire and science fiction, cryptography and angelology – reading with an eye for the tropes and techniques of speed can spring into visibility a complex three-dimensional world, with new urges, motives, needs, reactions to a whirlwind of change. (204)

What travel and narrative have in common is passage or conveyance and the temporal movements necessary to both; both are 'speedy messengers', to use the term that hovers in the title of Godwin's book. The title page has *The Speedy Messenger* in italics beneath the name of the putative author DOMINO GONSALES, separated by a line. There is no comma to make it clear that 'The Speedy Messenger' is being offered as an epithet or alternative designation for Gonsales, making it possible to read it as a gloss on or alternative title for the whole work. Nor is it quite clear what is meant to be being sent so speedily. Is a 'speedy messenger' one who sends messages at speed, or a messenger who is himself speedily moved? Is the sender despatched? Is the despatch itself a sender? Missive, missionary, missile?

For Godwin, the transport of bodies moves easily into the transmission of signs. Godwin specifies rather awkwardly that Gonsales and his faithful negro manservant must live apart on the island of Tenerife because of shortage of food, but really in order that their physical separation should bring about the need for communication, which the trained geese initially supply – 'I brought them to that passe, as that a white sheet being displayed unto them by Diego upon the side of a hill, they would carry from me unto him, Bread, flesh, or any other thing I list to send, and upon the like call returne unto mee againe'. Gonsales promises his reader that 'this Art containeth more mysteries than are to be set downe in a few words: Hereafter I will perhaps afford a discourse for it of purpose, assuring my selfe that it may prove exceedingly profitable unto mankind, being rightly used and well imployed: for that which a messenger cannot performe in many dayes, this may despatch in a peece of an houre' (Godwin 1638: 22). Godwin duly supplied this discourse, in the form of a Latin essay entitled 'Nuncius Inanimata', the 'lifeless messenger', that he added to the second edition of the work. John Wilkins, who knew of Godwin's work, also connected the idea of a passage to the moon with the approach to instantaneous transmission of messages (Wilkins 1638: 1641).

This colloquy or collocation between transportation as a physics (movement of mass through distance), and transmission as an informatics, that concerns itself with the near-instantaneous communication or co-appearance of signs or meanings – or, more specifically perhaps, an imaginary acceleration of the former into the latter – anticipates a contemporary condition in which speed makes for relations of pervasive proximity, rather than transmission between separated locations. In such a condition, physics in fact becomes information, and, as Vera Bühlmann and Ludger Hovestadt put it

> the behavior of processes media-ized by a form of technics, the processing unit of which is information, may not adequately be described through the physical-mechanics referential framework that is traditional to our understanding of technics [...] Because subjacent to the incipient *operable availability* of our gradually acquired techniques of symbolizing and operationalizing the relations, the interplay of which makes up our world, there changes, along with the technical substratum, likewise the substrate of our existence. This substrate reveals itself today as some sort of "symbolized physics" to which we can refer for the moment as *media-ized nature*. (Bühlmann and Hovestadt, 2014: 13)

Motion and Commotion

All machines have subjects and objects. They are devised and driven by their controllers, and they produce effects on their objects, whether these are sacks of grain, or persons who are carried through space. Ultimately, this is the sign and the outcome of the fact that all machines operate in time, that they are themselves moved through time through the motions they impart. Being so directed, machines all have direction.

But all machines also embody a counterfactual dream, of an undirected motion, of a motion that, because it moves itself, is not itself moved in any particular direction. The cosmos displays motion of this kind because its characteristic motions are circular, or perpetual, without end or beginning. As such they are both the production of God and, in a certain paradoxical sense independent of God, since they are, like God himself, self-moving 'automata'. Only a creature itself increate, it is imagined, could give rise to a Creation that seems similarly increate, or self-creating.

There are then two kinds of motion, onward and inward, extensive and intensive. The first is a motion that precedes and goes beyond the machine, or moves the machine itself through space, the second is a motion that the machine itself sufficiently is. The first kind of motion is emblematised in the piston rod, which converts circular motion into directed effects, or, alternatively, which converts alternating motion into motion in a certain direction, as in the wheels of a locomotive. In either case, however, there is conversion of one thing into another, which testifies to the arrow of time, and so must then be subject to the operations of the Second Law of Thermodynamics, meaning that a certain proportion of the energy expended must be lost.

The second kind of motion is that indicated in the vision of Ezekiel (1. 15-16):

> 15 Now as I beheld the living creatures, behold one wheel upon the earth by the living creatures, with his four faces.
>
> 16 The appearance of the wheels and their work was like unto the colour of a beryl: and they four had one likeness: and their appearance and their work was as it were a wheel in the middle of a wheel.

Ezekiel's singular 'wheel within a wheel', itself already an image of interior multiplication, multiplies into 'wheels within wheels'. The first expression was often used in the seventeenth and eighteenth centuries to signify reversal or alternation, as in a 1754 sermon by Ebenezer Erskine on 'The Wrath of Man Bounded by the Power of God': 'There is a wheel within a wheel, which will turn matters about so, as the wrath of man shall praise God, and advance his interest, instead of ruining it' (Erskine 1791: 711). Often, it is God, or the Prime Mover, who is the wheel within a wheel. Similarly, we read that 'the providence of God is like a wheel within a wheel' (Blackerby 1674: 180). But the wheel within a wheel also signifies eternity – static commotion rather than motion, that has no exterior principle – or rather the infinite reflections that the thought of eternity provokes, as in Abraham Caley's evocation:

> in *Eternity*, one deep calleth upon another; it is a wheel within a wheel, there is difficulty after difficulty, one mystery folded up in another; it is a great *Maze* or *Labyrinth* full of infinite *windings* and *turnings*: after all our searchings and indagations, we may well lose our selves, but can never retreat when our thoughts are seriously engaged in it, much less find any bound that may set limits to our meditations. (Caley 1683, 42)

At the heart of every propulsion machine, every machine that seems to propel not just its passengers but itself into pure passage, there may often be some impossible and terrifying thicket of self-consuming cogs and gears, some unreadable and dangerous mechanism, which can grind and crush rather than volatilise. This is magnificently imaged in the engine room scene of the film *Galaxy Quest*, in which Sigourney Weaver as actress Gwen DeMarco playing Lieutenant Tawny Madison and Tim Allen as Jason Nesmith playing Commander Peter Quincy Taggart have to negotiate their way through the engine room of their starship, in order to manually override the core shutdown sequence (or something). When Taggart proposes penetrating to the engine room by means of air ducts, DeMarco memorably protests: 'Ducts. Why is it always ducts?' Confronted eventually by the terrifying gnashing maw of the ship, she resumes her theme: 'What is this thing? I mean it serves no useful purpose for there to be a bunch of chompy, crushy things in the middle of a hallway. We shouldn't have to do this! It makes no logical sense! Why is this here?' When Nesmith explains 'Because it's on the television show', she replies 'Well forget it. I'm not doing it. This episode was badly written'. The gears are a wonderful image of the machinations of a plot that must be gone through simply because they are characters in a narrative (or, strictly speaking, a narrative within a narrative, a wheel within a wheel), and that is what narratives do to characters. Where the principle of the self-abolishing machine is that of self-consuming synthesis, in a speed which allows for no divisions to remain, the image of movement as pure milling is governed by the principle of minute division, perhaps recalling Sextus Empiricus's quotation from an unknown poet in his *Adversus Grammaticos*, Ὀψὲ θεῶν ἀλέουσι μύλοι, ἀλέουσι δὲ λεπτά, 'The millstones of the gods grind late, but they grind fine' (Sextus Empiricus 1998: 311).

There are two extreme, ideal forms of the space craft, which approximate to two kinds of body schema. In the first, the machine shrinks tightly round the body of the passenger, as in the ideal racing car. This kind of projectile finds its ultimate form in the human cannonball which, like Kafka's racing Indian, is reduced to the condition of pure projectile. The other body schema is the precise opposite. Where there is no space at all in the body-schema of the projectile, the interior of the classical space ship is almost all space – vast and scarcely even to be grasped. It is, paradoxically, a kind of labyrinth of openness, an unreadable space. It is, as in the sequence of *Alien* films, a mother-ship, a maternal, uterine expanse. As Kubrick's *2001: A Space Odyssey* seems to make clear, motion in such a space will be suspensive, rotary, orbital, despite the fact that the machine itself will be hurtling through space, like the Earth, at many thousands of miles an hour, or second.

In neither case will any machinery be apparent – or, at least, its becoming apparent, as in *Galaxy Quest*, appears as defamiliarisation. The machinery will have been replaced by the controls, consisting of screens and switches. This movement from action to communication, from effect to medium, is characteristic of many machines. The machine comes to be understood more and more as *operation* – as machinery for communicating with the machine.

Going To See

I have said that machines of transport work by converting commotion into motion. But the point of transport is often to recreate or make possible just such an agitation or commotion. When Boswell remonstrated with him that the Giant's Causeway was surely worth seeing, Dr Johnson replied 'worth seeing, yes; but not worth going to see' (Boswell 2008: 744). What is the transaction implied here? It is surely something like the promise that the expense and fatigue involved in travelling to some other place will produce an experience that will render it worthwhile. To jargon it, extension will make possible intensity. Put in mechanical terms, this might be seen as a reversal of the original transformation of commotion into directed motion: one travels to the Grand Canyon, Niagara Falls, the Taj Mahal in order to be transported, in order to experience an intensity of being there that is a kind of transport, a departure from oneself. Milton plays with these different kinds of transport in *Paradise Lost*, when Adam explains the new kind of delight that enters his world with the arrival of Eve:

> Thus have I told thee all my state, and brought
> My story to the sum of earthly bliss,
> Which I enjoy, and must confess to find
> In all things else delight indeed, but such
> As used or not, works in the mind no change,
> Nor vehement desire, these delicacies
> I mean of taste, sight, smell, herbs, fruits, and flow'rs,
> Walks, and the melody of birds; but here
> Far otherwise, transported I behold,
> Transported touch; here passion first I felt,
> Commotion strange, in all enjoyments else
> Superior and unmoved, here only weak. (Milton 1998: 298-9)

As Joe Moshenska has observed, transport and commotion are potent, even dangerous, words for Milton, both suggesting an incontinent loss of control (Moshenska 2014: 247). There is always the risk, in a dream of taking leave, of taking leave of one's senses. John Wilkins observes that dreams of mechanical flight have often been regarded as the kind of mental lightening or volatility characteristic of melancholy madness:

> Amongst other impediments of any strange invention or attempt, it is none of the meanest discouragements, that they are so generally derided by common opinion, being esteemed only as the dreams of a melancholy and distempered fantasy [...] therefore none will venture on any such vain attempt, as passing in the air [...] unless his brain be a little crazed with the humour of melancholy. (Wilkins 1680: 197)

The experience of modern travel is one in which the traveller seeks to minimise the cost in terms of physical inconvenience in order to maximise the gain – to increase the yield of excitement or commotion with respect to the outlay of motion. We seek to travel as pure intermediary between states, as opposed to being a mediator that is itself transformed by the transportation. Bruno Latour allegorises this difference in the example of a traveller making their lulled way across a landscape at hundreds of miles an hour on a jet aeroplane or bullet train as opposed to a traveller having to fight her (the possessive pronoun having the conventionally heroic form) way through impediments at ground level (Latour 1997: 174). The difference between transportation with and without transformation equates to the difference in the number of interruptions that one is likely to encounter (178).

Reducing the physical cost of speed to achieve the state of frictionless passage through space that turns space entirely into 'flying-time', itself takes a very long time to achieve. The cost is paid at earlier stages in various forms of what might be called motion sickness, a condition in which the body is disorientated by the replacement of motion by commotion. Speed intensifies the shock and cost of this commotion. The history of attempts to deal with the effect of blurred landscape, the rattles and shakes incident to 'railway spine', a condition thought to be caused by exposure to intense and continuous states of vibration, along with the development of pneumatic tyres and suspension systems, are all forms of this effort to reduce friction and turbulence, in the interests of motion without commotion. The result is a sort of princess-and-the-pea phenomenon whereby plane passengers can undergo extreme, stomach-churning stress at the kind of bumps that would be unnoticed if they were driving over a tarmac road. All of these can be thought of as a kind of nausea, a word which evokes, and is related both to the nautical and the noisy.

Perhaps all the stress of journeying is a version of sea sickness, that is the subject of a brilliant analysis in the lyrics to Irving Berlin's 'We Saw the Sea' in the 1936 musical *Follow the Fleet*: 'We joined the Navy to see the world / And what did we see? We saw the sea / We saw the Pacific and the Atlantic / But the Atlantic isn't romantic / And the Pacific isn't what it's cracked up to be'. You go to sea to see things, but there is nothing to see at sea except the grey monotony of the pestling waves, which are at once endlessly in motion and yet apparently entirely without variation. The song see-saws being seeing and doing – 'We joined the Navy to do or die / But we didn't do and we didn't die / We were much too busy looking at the ocean and the sky' – and between the monotony of the experience evoked and the ersatz excitement of the agitated rhymes and packed prosody of the lines. There is a kind of nausea that comes from the absence of motion, of a motion fed back into itself that leaves one simply 'all at sea': 'We never get seasick sailing the ocean / We don't object to feeling the motion / We're never seasick but we are awful sick of the sea'.

Tourism promises or promotes the fantasy of a kind of transport that is identical to television and, reciprocally, perhaps, a television that is supposed to be the same as transport, collapsing physics and information in just the same way as Godwin and Wilkins do in their writing of imaginary voyages and their ideas of telegraphic transportation of image and sign. But in touristic travel there is at least the promise or fantasy that things will act in a different fashion. The traveller feels entitled to remain intact in order to be speedily transported to a place where they will be able to experience a kind of stationary, concentrated, all-at-once transformation before different kinds of spectacle. It appears that the cost in commotion must always be defrayed at some point in the process. Even after the traveller has been cocooned from every kind of disturbance on the way to their destination, there is the payback in the temporal perturbation of jet-lag.

Teleportation

As it happens, there is an imaginary form of transport that seems to enact just these conditions, while also reproducing the paradoxes of immaterialisation associated with speed. Like many other technologies, teleportation began to be dreamed of among late nineteenth-century supernaturalists. Mediums borrowed the techniques of conjurers to produce 'apports', objects apparently materialised out of thin air, typically flowers, fruits and garments, but also sometimes objects like needles and grasshoppers. Madame Blavatsky turned this into a spiritual postal service, contriving to have astral letters from Mahatmas living in the Himalayas, including her primary master, Koot Hoomi, supplied to her and left in unexpected places to be discovered by fellow-theosophists. A report for the Society of Psychical Research led by Richard Hodgson revealed the connivance of her maid, Emma Colomb, who would post the letters through a crack in the ceiling, and the existence of a shrine with a false back which allowed letters to be deposited overnight (Melton 2001: 2.968). It was not long before mediums started to reproduce the illusions practised by conjurors to effect 'teleportations' of persons. Edmund Gurney's *Phantasms of the Living* (1886) recorded many cases of telepathic transfer of images, voices, odours, and impressions. These are not presented as actual examples of physical transportation, but they encouraged a revival of the belief in various kinds of astral projection. Most of those who evoked this phenomenon kept things safely metaphorical, by reference to movement between various kinds of 'plane'. As usual, though, Madame Blavatsky provides the comic-grotesque link between the spatial and the metaphorical, for her astral travel is also astronomical, involving the transit between planets.

Teletechnologies from the telegraph onwards have attempted to breach the barrier that separates the body from its signs or traces. The telephone suggested to many that some physical essence might be transferred through or along with the sound of the voice – the possibilities of telephonic osculation were being mooted as early as 1876 (Connor 2000: 382). Morse telegraphers had their own distinctive rhythms or bodily signatures, transmitted as a kind of identifying noise with the message, and enabling them to be recognised. One of the ways in which it has been possible to transmit messages without the possibility of interception has been to make the body of the messenger identical with the message; messages were sometimes tattooed on to the scalps of messengers, a practice that seems cybernetically equivalent to systems like iris recognition which make the carrier or channel of the code the message itself. The principle behind this is that a message that is unknown to its carrier is much more secure than a message which is, precisely because it cannot be sequestered from him. The ideal here is that the transmitter might be transmitted along with their message.

David Trotter has described the ways in which, from the late nineteenth century onwards, the experience of being 'in transit', in trains, ships and, later on, in aeroplanes, became ever more common, and, despite the prodigious increases in the speed of travel, seemed to occupy ever more time, time which needed actively to be filled, typically through reading, and perhaps more and more often reading of the kind of travel literature that began to be produced in large quantities. As the traveller took a less and less active role in the process of being conveyed, the kinetic and the encoded seemed to come closer and closer together:

> The purpose of mass-transit systems is to connect. The more closely the traveler resembles a "living parcel," an object to be laid in a rack rather than a subject prone to look out of windows, the more efficiently she or he will be conveyed from one place to another, and with the least possible attendant misery. To put it in the yet starker terms appropriate to twentieth-century mass-transit systems, the more closely he or she resembles a message transmitted from one place to another, the better. (Trotter 2013: 220)

The very wearisomeness of these oxymoronic ordeals, in which simply doing nothing for long periods seemed to produce as much fatigue as if one had been spurring a horse all the way, must have encouraged the idea that, if only the body could be fully converted into a message, these long periods of transit could be cut out. The most familiar fantasies of teleportation involve the idea that if the body as such could be translated holus-bolus into information, it could be transported as swiftly and losslessly as information apparently can. Transported in this way, the body would become the speedy messenger of itself, both passenger and vehicle. If teleportation is the most familiar projection of the idea of bodily projection, a version of this fantasy is that of 3-D printing, which allows for a form to be copied and duplicated at a distance from its original. This is a mechanised form of the Incarnation, in which Christ is commonly thought to assume or take on fleshly form, or the manifestations of spirits in the nineteenth-century séance room, where a certain quantity of ectoplasm or, as it was often more tellingly called, teleplasm, needed to be gathered from the collective credulity of those present to build the manifestation. One of the reassuringly scientific-sounding names given to materialisation among spiritualists and psychic researchers was 'teleplastics' (Schrenk-Notzing 1923). Accounts of séances marvel at the capacity of mediums to form 'moulds' of plaster or paraffin wax, which could be retained after the séance, the spirit hands having dematerialised from within the artificial 'glove' they had formed. Charles Richet described one such materialisation in his *Thirty Years of Psychical Research* of 1923:

> I have [...] been able to see the first lineaments of materializations as they were formed. A kind of liquid or pasty jelly emerges from the mouth or the breast of Marthe which organizes itself by degrees, acquiring the shape of a face or a limb. Under very good conditions of visibility, I have seen this paste spread on my knee, and slowly take form so as to show the rudiment of the radius, the cubitus, or metacarpal bone whose increasing pressure I could feel on my knee. (Richet 1923: 469)

Richet provided the rationale for these appearances by analogy with other forms of energy 'projection':

> Materialization is a mechanical projection: we already know the projection of light, of heat, and of electricity; it is not a very long step to think that a projection of mechanical energy may be possible. The remarkable demonstrations of Einstein show how close mechanical and luminous energy are to one another. (468)

The most common materialisations were of human forms, but there were also locks of hair and animals, including, frogs, dogs, birds and weasels. Intermediate and ill-defined forms that were themselves in transition between conditions were sometimes produced: for example, in a séance held in November 1920 with the Polish medium Franek Kluski, 'a strange creature intermediate between ape and man' named the '*Pithecanthropus*'. The Pithecanthropus was described as 'having the stature of a man, with a simian face but a high, straight forehead, face and body hairy, with long arms and very strong hands. Seems emotional, takes the hands of the Sitters and licks them like a dog' (Geley 1927: 258). For obvious practical reasons, detached and mobile body parts were also very common in séances, for example the form of a child's arm that extruded itself from the elbow of the medium Eva Carrière, known as 'Eva C.':

> We then suggested that Eva should produce a well-developed forearm with hand. The medium, whose hands were constantly controlled, evidently made strong efforts to carry out the

> suggestion. She made deep and audible respirations, and groaned and whimpered. At last we saw, on the inner side or her left arm, and starting from the elbow, the outlines of a left forearm, gradually becoming more distinct. A hand and fingers were formed in a rudimentary and imperfect way at first. But, before our eyes, this semi-liquid substance, endowed with some kind of animal life, changed its appearance, until it assumed the form of a correctly-drawn left hand, somewhat smaller than the medium's arm, and showing all the imperfections already mentioned as regards modelling, muscular development, detailed structure, and nails. (Schrenk-Notzing 1923: 74-5)

The 'vitalised structure' then detached itself from the elbow and floated up to join itself to her shoulder (75). There were also some very ambitious inorganic materialisations, like the form of a three-dimensional teleplasmic sailing ship materialised in the Winnipeg séance circle in 1903 by the spirits known as 'John King' and 'Walter'. It was assumed by T. Glenn Hamilton, who wrote a report of the materialisation in 1933, that it must be the work of spirit guides rather than of the medium:

> No matter how great we may conceive the unknown powers of the human organism to be, we cannot conceive of it giving rise to an objective mass showing purposive mechanistic construction such as that disclosed in the ship teleplasm of June 4th [1903]. We are forced to conclude that the supernormal personalities in this case (by some means as yet unknown to us) so manipulated or otherwise the primary materialising substance after it had left the body of the medium, or was otherwise brought into its objective state, as to cause it to represent the idea which they, the unseen directors, had in view, namely the idea of a sailing ship. (quoted Melton 2001: 2.988)

Unfortunately, there was a little hesitation in giving the signal to take a flash photograph, with the result that the form of the ship was imperfect – or, as Hamilton reports it, 'came into port badly damaged' (2.988). Imperfection could be the proof of the telepathic transit of forms, and so it was possible for some scrambling of the message to take place in séance projections. The medium Elizabeth D'Esperance described a manifestation in which a materialised form standing beside her was photographed with her face, while her seated form bore the face of another:

> Now comes the strangest part of this strange experiment. The photographic plate was carefully developed and a print made, which revealed a most astonishing fact. The materialised form, well in focus, was clad in white, flowing garments. The hair

was hanging loosely over the shoulder, which, like the arms, were without covering. The figure might have been that of a stranger, but the features were unmistakably mine. Never has a photograph shown a better likeness. On a chair beside it and a little behind, was a figure clad in my dress, the black bands on the wrist, and the tape round the waist showing themselves clearly and intact, but the face was that of a stranger, who seemed to be regarding the proceedings with great complacency and satisfaction. Needless to say, we looked at this extraordinary photograph with something like petrifaction. (989)

Even the medium's control, the spirit known as 'Walter'. was a little arch: 'We were utterly at a loss to understand its meaning, and no explanation was forthcoming, except a rueful remark from Walter, who when questioned replied that "Things did get considerably mixed up"' (2.989).

Everywhere But Back To Me

It became clear early on in electronic telecommunications that the chances of interference when any kind of message was transmitted were high, and very hard indeed to reduce to nothing, though the principle that there is no entirely noise-free channel would have to wait until the work of Claude Shannon and other early information theorists to be formulated theoretically. Experience would also have shown how easy it is for physical transportations to include stowaways and unexpected passengers, whether in the form of animals, organisms or infections. The possibility of these superfluous entities getting compounded with the principal subjects of the transmission is in the offing from the beginning of thinking about the kind of transmission dreamed of in teleportation. It is, in particular, the animating principle behind the most well-known story of this kind, 'The Fly'. The first version of the story was written by George Langelaan and published in *Playboy* in 1957. It was released as a film the following year, and followed by a number of sequels and a remake by David Cronenberg in 1986, which has also spawned sequels.

Langelaan's story establishes from the outset the links between transportation and the transmission of information. The story with a telephone call made to the narrator by his sister-in-law Hélène Delambre, to tell him that she has just killed his brother André by crushing his head under a steam-hammer. The narrator's irritation at the easy intrusion that can be effected by the telephone looks forward to the intrusion of the fly into the telephone booth which André Delambre has adapted to act as his teleportation chamber:

> Telephones and telephone bells have always made me uneasy. Years ago, when they were mostly wall fixtures, I disliked them,

> but nowadays, when they are planted in every nook and corner, they are a downright intrusion [...] At the office, the sudden ringing of the telephone annoys me. It means that, no matter what I am doing, in spite of the switchboard operator, in spite of my secretary, in spite of doors and walls, some unknown person is coming into the room and onto my desk to talk right into my very ear, confidentially – whether I like it or not. (Langelaan 1964: 7)

The fly is both the synecdoche and the emblem of the incidental. Always being liable to be at hand, flies are a kind of essential accident, the sign of the insignificant and the metaphor of metonymity, since a metonym involves a substitution of some associated or adjoining quality, something accidental rather than essential. Flies commute rapidly between a human kind of speed – they walk and crawl, often over our persons – and a speed of movement and response that is much greater than ours. Because they are always on their way, a way that seems to consist of pure passage rather than purposive movement towards a goal, they are liable to get in the way. It is for this reason that flies have often been thought of, not just as interlopers or intermediaries, but also as spies or overlookers – the 'fly on the wall'. Flies also suggest perceptual interchange in a way that other similar creatures usually do not. How we might be an object of perception for a beetle or a slug is not usually a concern for us, but the strong association of the fly with vision means that it seems to contain the possibility of a drastic inversion of perspective and value, in which it is we who are the accidental or background phenomenon.

Bruce Clarke suggests that 'the teleporter technology of the *Fly*s of the '50s remains tethered to a classical-materialist mishmash of physics unaffected by electromagnetic-field theory, relativity, or the cybernetic coupling of material-energetic and informatic ensembles' (Clarke 2002: 179). This may perhaps be true of the way in which the technology is explained, but its function within the narrative suggests that transport and transmission are more closely meshed together. The fly in Langelaan's story is first transcribed then transported – that is, it is sampled then reassembled in another place. The sorting remains latent, making the fly merely available to be recombined in any number of ways. It is as though, for the time being, the fly's components and their relationships have just been numbered, or even named. And yet the sorting must already in some sense be a transporting. What if the fly had been simply sorted, then resorted back into exactly the same genetic and molecular configuration without being shifted to another place? The fly would in fact have been exported away from itself then returned to itself – a transport consisting entirely of a report, a word we mostly apply to signs rather than portable objects, though being turned into a sign makes an object eminently portable. More than this, might it not be that the act of sorting must be seen as a deporting-reporting dyad, whether or not the fly is transported to some other location to be reassembled? To sort the fly is to deport it from itself in order to report it to itself. The fly, so often identified with the parasite, the interloper, or the interrupting accident that both disorders and reorders, would here be subjected to the process of which it is commonly, as Maxwell's neat-fingered indexing demon, the subject. It is itself interrupted, into its own, essentially interruptive, being. The act of informing the fly, turning its physical form into information, is equivalent to a physics.

This alternation between physics and information is also a feature of the story that, as we say, conveys the fly to us as readers. 'The Fly' is constructed as a detective story, in which two different temporal sequences intersect. There is first of all the order of the events in, so-to-speak, real time – André Delambre's discovery of the matter-transporting method, his accident, his efforts to reverse it with Hélène, the assisted suicide, arrest, and its aftermath. There is also the order of the narration, which begins in the middle of this sequence. The story employs the traditional device of an interior narrative in order to synchronise these two sequences, in the form of the manuscript narrative which Hélène produces for her brother-in-law and which is then incorporated into his narrative. Narrative movement through reading-time is thereby turned into an object in space – the 'thick wad of closely written pages' (Langelaan 1964: 20-1),which our narrator, Francois, incorporates into his narrative for us to read, and which is subsequently handed to the police detective, Commissaire Charas, to read in his turn, after which 'he carefully folded Hélène's manuscript, slipped it into the brown envelope, and put it into the fire' (41).

Form and content intersect and interfere with each other in the same way as the insect (its name really meaning a kind of intersection, or cutting across) interferes with the self-transmission of André Delambre across the floor of his laboratory. Some of these interferences seem simply arbitrary, such as the fact that the efforts to get Hélène to reveal the truth about how and why she has killed her husband seem to recapitulate the nature of his death:

> I suddenly realized that here was the opening I had been searching for and perhaps even the possibility of striking a great blow, a blow perhaps powerful enough to shatter her stonewall defence, be it sane or insane.
> Watching her carefully, I replied:
> 'I don't really know, Hélène; but the fly you were looking for was in my study this morning.'
> No doubt about it I had struck a shattering blow. She swung her head round with such force that I heard the bones crack in her neck. (17-18)

There is also the even more arbitrary entomological reappearance in the narrator's joke about the difference between brandy and whisky: 'I handed him his brandy and mixed myself what he called "crushed-bug juice in soda water" – his appreciation of whisky' (41).

But the real correspondence between the narrated and the narrating is structural. As with many detective stories, the story is not so much an elaboration as an undoing of an originary enigma or complication. Detective stories often depend upon what is called a *dénouement*. In 'The Fly' that principle appears to be made literal, for the composition of the story is matched by the decomposition of its elements – both the traditional 'unravelling' of a mystery and the reduction of its elements to the unspeakable, nauseous mess of André's body, or the ash to which Hélène's manuscript is reduced. Nowadays, the word *ravel* is rarely found unaccompanied by its negative prefix but it was common in the early seventeenth century, by which time it had attained the condition of one of Freud's 'antithetical words' (Freud 1953-74: 11.153-62). As with drawing a curtain, which can mean both to draw it and to undraw it, so to ravel could mean both to fray, knot up, and confuse, and also, as 'ravel out', to explicate or make plain, or as 'ravel back', as in a 1661 observation by John Fountain that 'men ravel back to childhood, when / They cease to be thy Children, sacred Vertue' (Fountain 1661: 28), to reduce or return to an original condition of order or simplicity. In Macbeth's reference to 'sleep that knits up the ravelled sleave of care' (Shakespeare 2015: 181), Shakespeare may be relying on the hearer being able to thread a little of the ordering sense of ravel into the idea of sleep. What has to be unravelled in 'The Fly' is the story of an attempt to undo an unravelling. Its efforts are expended in *sauter pour mieux reculer*.

The reversal is something that affects and refers to the words which are the medium of the story. The first hint at what might go amiss in transmission is the fact that the words Made in Japan on the bottom of the 'typically British souvenir' come out as napaJ ni edam (Langelaan 1964: 23). Capital letters are a physical sign of the decomposition of physical form in the story. The first and most palpable sign of André's transformation is that he can no longer write by hand, and must leave his wife typed notes instead. The typewriter signifies the loss of the fluent continuity of manuscript, since type reduces language to permutable, but identical elements. The narration plays with these alternative conditions, giving us, in the printed form of the story, the narrator's transcription of Hélène's manuscript account (has he made a copy, in which case why make so much of the fact that the Commissaire destroys the manuscript after reading it, or has he committed it to memory rather than the fire?), which incorporates reproductions of André's typewritten notes. Are we to assume that Hélène has taken the trouble to write these notes out in capital letters, or is this an editorial improvement offered by her brother as narrator, like a fly applying some readerly ointment?

Everything goes wrong of course because of the primal disturbance effected by André's experiments with the disintegration and reintegration of matter. André has projected a world in which, with transportation transformed into a kind of transcription, there will be no need for anybody to undergo physical movement:

> André considered his discovery as perhaps the most important since that of the wheel sawn off the end of a tree trunk. He reckoned that the transmission of matter by instantaneous 'disintegration-reintegration' would completely change life as we had known it so far. It would mean the end of all means of transport, not only of goods including food, but also of human beings. André, the practical scientist who never allowed theories or daydreams to get the better of him, already foresaw the time when there would no longer be any airplanes, ships, trains or cars and, therefore, no longer any roads or railway lines, ports, airports or stations. All that would be replaced by matter-transmitting and receiving stations throughout the world. Travellers and goods would be placed in special cabins and, at a given signal, would simply disappear and reappear almost immediately at the chosen receiving station. (21)

But all of this depends upon the reduction of noise to zero, upon the possibility of the absolute reproduction of a code without change or error. Instead of the orderly transmission of signs, goods and persons of which André has dreamt, the teleportation apparatus seems to have produced the wreckage of a sort of primal Fall:

> My first impression was that some sort of hurricane must have blown out of the receiving booth. Papers were scattered in every direction, a whole row of test tubes lay smashed in a corner, chairs and stools were upset and one of the window curtains hung half torn from its bent rod. In a large enamel basin on the floor a heap of burned documents was still smouldering. (30)

One imagines that Tom Stoppard might have encountered this story or the film version of it. When he was asked to write, rehearse and produce a play in a week for the BBC series *Eleventh Hour* in 1975, the result was *The Boundary*, written with Clive Exton, in which a lexicographer (naturally named Johnson) enters his study to find his notes and papers in disarray and his wife apparently dead underneath them. The disordering that is the primal engine of plot extends not only to the events but also to the language of the play, which becomes as *distrait* as the disassembled dictionary. Entering the room, Johnson first tries to phone for assistance using the television and then a telescope. Locating the telephone at last, he announces 'Now look here my name is Johnson. I seem to have been bungled' (Stoppard 1991: 2).

The story also seems to communicate intriguingly with some features of Langelaan's own life. Langelaan had been a secret agent during the Second

World War. The critic Albert Guérard recalled working with him, searching for covert anarchist forces in France during 1944. The situation was one in which lines of authority and control were extremely uncertain. With victory over the Germans looking likely, British and Americans became concerned at the possibility of a take-over of power by anarchists or communists, or even the 'diehard collaborationists' (Guérard 1969: 444) who believed that the real enemy was Russia, not Germany. He records that Langelaan had been asked 'to explore the possibility of our smuggling André le Troquer into Paris before the liberation, to obviate a seizure of power' (443). Guérard remembers Langelaan as at once flamboyant, risk-taking and terrifyingly calm. Guérard and Langelaan were on the track of a shadowy army of 25,000 Spanish anarchists. The bathetic climax of the story comes with an encounter with Langelaan 'in his Paris office (my enigmatic British agent sadly translated to a typewriter, as courageous and brilliant as James Bond, but out of place in civilian clothes)' (452). Langelaan confirmed that he had in fact discovered the anarchists in Vernet-les-Bains:

> 'What were they doing?'
> 'They were doing close order drill.'
> 'What sort of weapons?'
> He hesitated, took a long meditative pull at his pipe.
> 'There was a small factory for brooms near there.'
> 'Brooms?'
> 'They were drilling with broomsticks.'
> 'And there were how many – out of that twenty-five thousand we were sent to find?'
> 'One hundred and fifty-six.' (453)

Langelaan published two volumes of reminiscences of his war years, *The Masks of War* and *Knights of the Floating Silk*. The first of these begins with his experiences as an intelligence officer during the retreat to Dunkirk in 1940. It spends a great deal of time reflecting on the relationship between the physical movement of equipment and people, military and civilian, and the not-quite parallel movements of rumour, with one chapter centring on rumours of hungry tigers being parachuted into Northern France by the Germans, and the necessity of frisking nuns in search of the shaving gear that would identify them as male spies. Langelaan describes an exercise that has parallels with the efforts of André Delambre to transmit himself back to his original condition:

> As Field Security policemen we were supposed, not only to investigate rumours, but to trace them right down to their source, of course. No one, to my knowledge, has ever performed this stupendous feat. Just for the fun of it I one day tried to trace a rumour which I had myself started; my investigation led me

everywhere but back to me. But that is another story, as Kipling would have said. (Langelaan 1959: 15-16)

Machines of transport are always embodiments of counterfactual dreams of overcoming space and distance. Imaginary forms of transportation involve the tension between the solid and visible *being-there* of machinery and the dissolution of place and the body involved in the ideal of absolute vehiculation enacted in Kafka's dissolving rider. Machines and contrivances of all kinds assist with our dream of travelling at the speed, and even by means, of thought, just as we travel forward into the future or the past in reverie and reminiscence. And yet we aim not only to depart, but also to arrive and, as we tellingly say, 'all in one piece', as though to recognise the threat of dissolution that accompanies every desire for transport. Our imaginary machines are ways of accommodating ourselves to this duality of dwelling and passage, being there and the being-nowhere of going-there.

3

Dream Machines

The feelings we have for and in dreams are often mediated by the objects of which we dream (whether asleep or awake, and so taking dreaming in its largest sense), as well as the sorts of objects that dreams themselves may be taken to be. Those objects are sometimes, I think, mechanical in form and function.

A machine is a thing. It is on the object side of things. Yet a machine is an anomalous kind of thing, an object that seems to exceed its objecthood in certain ways, through its quality of being automatic, of moving itself. Through its capacity for motion, a motor is an object that seems to be moving across into the condition of a subject, or quasi-subject. Machines do work for us, and so a machine is always a kind of substitute for a subject. And yet, as Michel Serres says, subjectivity is already substitution: 'one must think of the subject as the potential for substitution. What does substitution mean? It is the same word as substance' (Serres 2014: 88; my translation). A machine stands in vicariously for that which has, and so itself is, the potential for vicariance itself. Subjects are not machines, because machines are objects; but they can imagine themselves as machines, as imaginary machines.

A machine transmits force. It has motion rather than emotion. But what if the force transmitted by a machine is the force of fantasy, or what may come to the same thing, the fantasy of force? What kind of transport does such a machinery effect? The word transport moves between different registers of transmission – the physical movement of objects or energies and the movement of feeling, the feeling, for instance, of being, as we say, moved.

This chapter concerns different kinds of dream machines – machines of which we dream, machines for generating and controlling dreams, and the machinery we imagine dreaming itself to be. So it is also concerned with the substitutive relationship between fantasy and machinery. This is always a question of feeling, because it is always a question of force. The force in question may be wholly imaginary; but the fantasy of force always exercises a real force, the force of fantasy.

There are many things to which the term 'dream machine' has been applied. One of the commonest applications of the phrase nowadays may be to a particularly magnificent kind of car or motorcycle, sometimes a racing car, and sometimes one that has been modified and elaborated well beyond ordinary specifications. A 'dream machine' here may not mean much more than simply the machine of your dreams, or beyond your wildest ones. But the term 'dream machine' is also commonly used for the Hollywood film industry and, by extension, the culture industry more generally. The dream machine of contemporary entertainment is not just what you dream of, it is what you dream with, since it is the machinery of your dreams. In July 2015 Google made available the source code for its *Deep Dream* software, which builds on the company's face-recognition algorithms to detect patterns in photographs and then enhance and amplify them. As in so many other examples of oneirurgic machines, the result is not so much the production of dreams, as the reproduction of a standardised notion of what a dream is. In the case of the Google software, it looks as though we are to imagine dreams always consist of the fractal repetition of eyes.

In M. Mitchell Waldrop's *The Dream Machine: J.C.R. Licklider and the Revolution That Made Computing Personal* (2002), the phrase 'dream machine' is applied to the personal computer, and the role in its development of a psychologist who was absorbed in 'the challenge of deciphering that ultimate gadget, the brain' (Waldrop 2002: 12). Waldrop's usage refers back to Ted Nelson's *Computer Lib/Dream Machines* (1974), a manifesto for personal computing that took the form of two books bound back to back and printed at 180° to each other. One half of the book, *Computer Lib*, is an attack on the secretiveness and centralisation of computing technology, technologists, and technicians. The other, *Dream Machines*, is an evocation of the many uses which the computer would have once it became liberated from centralised technical and bureaucratic control. Nelson sees computers not as apparatus, but as media, of a peculiarly ubiquitous and liquid kind: 'we live in media, as fish live in water [...] But today, at this moment, we can and must design the media, design the molecules of our new water' (Nelson 2003: 306). The two conjoined halves of the book are expressive of the idea that the mental and the technical may be fused and of the ways in which this may occur: 'To work at a highly responsive computer display screen, for instance, can be deeply exciting, like flying an airplane through a canyon, or talking to somebody brilliant' (306).

But in the 1960s another kind of dream machine appeared. In his book *The Living Brain* (1953), W. Grey Walter provided a description of the operations of the electro-encephalograph (EEG), which implies parallels between the machine and the neural machinery it is designed to register:

> The equipment used today for studying brain activity contains many electronic parts and devices which were developed for radar apparatus during the war. An EEG recorder usually has over a hundred tubes, resistances, condensers and so forth, with many rows of calibrating and operating knobs and switches. Its

> formidable and intricate appearance not infrequently prompts the uninitiated to ask whether such a display of ingenuity is really necessary. But if we consider the complexity of the object which it is designed and constructed to examine, the most elaborate EEC equipment can only be regarded as comparatively simple in design – and extremely coarse and clumsy in construction. (Walter 1963: 87)

The most striking discovery of EEG investigation, pioneered by Hans Berger in 1924, was the existence of regular rhythms of electrical discharge in the brain, with epilepsy being the most characteristic signature of disturbed rhythms. In the chapter of his book entitled 'Revelation by Flicker', Walter described the ways in which the movement from brain to detecting apparatus might be reversed, through the subjection of the brain to rhythmic stimulus through various kinds of flicker apparatus, of which the most familiar was a spinning wheel, perforated to produce pulses of light at regular intervals. Walter's purpose was to create a kind of interchange, or flicker-effect between two kinds of flicker, that of the machine and that of the brain that might be induced to respond to it, in what he described as an 'instance of investigation by the communication engineer's method of the Black Box: without ever looking into the box a good deal can be learned about what is going on inside by checking incoming signals against outgoing signals' (104). Walter speculated explicitly about the interchange between the external and the internal machines:

> When flicker is used, the display given by the toposcope comes near to being a moving picture of a mind possessed in quite another way. The correspondence between the extent and complexity of the evoked responses on the one hand, and the hallucinations of the subject on the other, is striking. The more vivid and bizarre the experience of the subject, the farther from the visual areas are the evoked responses, and the more peculiar their form and geometry. (111)

In 1970, Walter had a motor-scooter accident which caused him extensive brain damage and following which he was unconscious for two weeks. In an essay entitled 'My Miracle', he details his efforts to develop a learning programme to enable him to learn to think again, in a remarkable exercise of imaginary self-reconstruction. Walter is known not just for neurological research but also for research in robotics and artificial intelligence. Indeed, the contributions he made to neurology during the 1930s arose largely from his expertise in electronics. As he says in the essay he wrote about the extremely unlucky accident, his electronic skill 'was a very lucky accident since the study of brain dynamics started as a combination of electro-technology and physiology' (Walter 1972: 49). 'My Miracle' joins the act of writing to technical and mechanical processes, as Walter describes his work on himself, in cooperation with his colleagues at the Burden Institute, in order to facilitate his own recovery to the point (which unfortunately seems never to have been reached) where he could take up his work again:

> I was regaining my original mentality but my remaining difficulty in finding an easy way to solutions alarmed me – I couldn't at that time see my way to cultivate creativity. So I decided to make myself accessible to my professional friends so that they could confide in me and share their dreams as well as their problems. That is what I enjoy most and I don't think of it as 'work', although it needs training and practice like an elaborate game or sport. (49)

Meanwhile, Walter's research had been taken in other directions. While on a bus travelling through a long avenue of trees in the South of France, the artist Bryon Gysin had an experience of flicker-induced hallucinations, as he records in a diary entry of 21 December 1958:

> We ran through a long avenue of trees and I closed my eyes against the setting sun. An overwhelming flood of intensely bright patterns in supernatural colours exploded behind my eyelids: a multi-dimensional kaleidoscope whirling out through space. I was swept out of time. I was out in a world of infinite number. The vision stopped abruptly as we left the trees. Was that a vision? What happened to me? (Cecil 1996: 5)

Just over a year later, Ian Sommerville, a young computer technician who lived and worked closely with Gysin and William Burroughs, wrote to Gysin from Cambridge on 15 February 1960 explaining that he had been encouraged by what he read in Walter's *The Living Brain* to construct a device which could be used to induce the intense visual sensations Gysin had experienced. It consisted simply of a cylinder of card with perforations cut in it mounted on the turntable of a record player. A 100-watt light bulb was suspended in the middle of the cylinder, which, when rotated at 45 or 78 rpm, produced a regular pattern of flickers. Gysin modified it by adding his own paintings to the cylinder and took out a patent for his 'Procedure and apparatus for the production of active visual sensations' to which he gave the compressed name 'Dreamachine' (6). This device reproduced the flicker effect of the stroboscope used in the laboratory, though it also reproduced a problem which was not overcome until the development of electronic stroboscopes after the Second World War, namely that as flicker speeds increased, the duration of the flashes decreased (ter Meulen et. al. 2009: 317). One of the principal uses of the stroboscope had been to test the regularity of movement in conveyor-belts and record turntables, as well as to 'freeze' the vibrating folds of the larynx to allow inspection (Woo 2010: 4).

The belief is that the Dreamachine induces and amplifies neural oscillations, in the range 7.5-12.5hz, in the brain of somebody sitting in front of it with their eyes closed, inducing visual hallucinations. Alpha waves are associated with experiences of relaxation or meditation when the eyes are closed, and also (in a different form) with the state of REM sleep associated with dreaming. This suggests that alpha waves may be particularly associated with the state of waking or ambivalent dream, a dream that can be superintended and, as it may appear, mechanically regulated.

The production of hallucinations or visions through regularly flickering light had been reported at intervals before this. David Brewster claimed in 1834 that

> a remarkable structure may be exhibited at any time, and whether the eyes are open or shut, by subjecting the retina to the action of successive impulses of light. If, when we are walking beside a high iron railing, we direct the closed eye to the sun so that his light shall be successively interrupted by the iron rails, a structure resembling a kaleidoscopic pattern, having the *foramen centrale* in its centre, will be rudely seen. The pattern is not formed in distinct lines, but by patches of reddish light of different degrees of intensity. When the sun's rays are powerful, and when their successive action has been kept up for a short time, the whole field of vision is filled with a brilliant pattern, as if it consisted of the brightest tartan, composed of red and green squares of dazzling brightness. (Brewster 1834: 241)

Brewster found that a similar effect was produced by looking at the sun while moving the distended fingers of the hand from left to right, and also while looking through the slits in a phenakistoscope, a popular apparatus very similar to Sommerville's device (242). He assumed that what was being seen was the reticular structure of the retina itself. Others took these visual effects to be spiritual visions. Genesis P.Orridge makes much of the experience:

> The Dreamachine can quite literally invoke. It can call out that same blue light mentioned in high Egyptian magic and in Sufi texts. The energy Dervish Dance calls out, and which is received and then earthed by the pointing of the hands up and down from and to the Earth, is this same Light/Energy. (quoted Cecil 1996: 19)

Gysin discovered that adding the sound of his breath to the experience meant that 'I was more able to control the visual images I was receiving by the variation of breath, modulation, frequency and depth. I could hold, freeze-frame, loosen and shatter images; though I could not, nor did I wish to, control their content' (19). Ian McFadyen describes the Dreamachine as 'a form of psychic cinema, a magical machine triggering the projection of inner visions through electrical rhythms in the brain' (22). In the 2007 documentary film *FLicKeR*, Nik Sheehan describes the Dreamachine as 'a kind of portal into the time-space continuum [as though other things, indeed every conceivable other thing, were not]. It opens a window into a magical universe, a very real place inside all of our heads' (Sheehan 2007). Apparatus and apparition are closely entwined in kinetic and optical developments during the nineteenth century. The stroboscopic effects so characteristic and beloved of 'psychedelic' art of the 1960s, and its fantasies about the power of fantasy, have their origin in the phenakistoscope. The optical toys marketed during the nineteenth century, the thaumatrope, the phantasmascope, and the zoetrope, preceded the cinema, exploiting the so-called 'persistence of vision' principle to create the illusion of motion. One can see the Dreamachine as a deliberate attempt to break up the feeling of continuity provided by this kind of device. It was thought of as an optical equivalent to the work of discursive jamming or interference allegedly effected by the Burroughsian cut-up. Though conceived of as a sort of parallel to the standardising dream machine of the mass media, it was also intended as a disruption of its anaesthetising uniformity. Indeed, one can see the stroboscope as an illustration of the very function it is supposed to perform, of jamming or putting a spoke in the wheel (a *Spaniard in the Works*, to quote the title of John Lennon's 1965 whimsy) of ordinary mass-produced experience. It provided a sort of amplification of the interference effects common in Western films, in which wagon wheels appear to be moving backwards. In place of shared images, projected outwards into the external world, the Dreamachine was supposed to work at the level of the optical, or even of cognition, to create a kind of individualistic dream cinema, which was at once predictable (it worked mechanically) and unpredictable (for the nature of the 'dreams' could not be prescribed). As Dave Geiger puts it

> You imagine a nation of people glued to their television sets, sitting in their living rooms, Mum, Dad, kids, dog, cat, all in this blue-grey light, bathed in this, like a nation of automatons. Suddenly you have an alternative to that. The alternative is a kind of beautiful device that moves and allows each of the people in that room to have a completely different experience. There's no central authority projecting this from a studio somewhere, but rather each of these people inventing their own scripts and their own films. This was the ultimate way to defeat control. (Sheehan 2007)

It is also, one might think, the ultimate way to block the possibility of any kind of directed or concerted action. As Marianne Faithfull remarks, 'It is like a wonderful idealistic idea. But you know it's never going to fly. People unfortunately prefer television' (Sheehan 2007). Yet in many respects the Dreamachine may be regarded not as the antidote to television but as its apotheosis, inducing passive absorption in an oneiric theatre in which it is unclear how we are to distinguish the inside and the outside, the viewing and the viewed.

The countercultural company kept by the Dreamachine also made for close associations with the altered states of consciousness sought, more or less programmatically, by Burroughs and others through psychoactive drugs. Drugs may seem more allied to natural or organic modes of transforming one's mental experience than machines, but there was in fact a strong alignment between drugs and technology. As Evelyn Fox Keller has observed, 'one might argue that psycho-pharmaceuticals have been more effective in persuading people of their essentially mechanistic and physical-chemical nature than all of modern science put together' (Keller 2007: 357).

At various points in human history, techniques have been applied not only to interpret dreams, but also to produce them. The process of procuring or governing dreams, setting the dreamwork to work, as we might say, is often known as incubation, after a practice that was common in Greek and early Christian times. Incubation is from *cubare* and *cumbere*, to recline or bear down upon – a root that gives us the cubicle, incumbency, succumbing and both the incubus and succubus. It seems likely that there were prescribed ritual procedures which had to be followed in order to provoke either the healing intervention of the God himself (commonly Asclepius) in a dream, or a dream in which advice was given as to the healing regimen to be followed. Although little evidence has survived concerning the detail of these procedures, there are records of some of the actions to be performed by those seeking incubatory cures at the oracle of the chthonian deity Amphiaraos at Oropos. Pausanias relates that it was required that the patient sacrifice a black ram and sleep on its spread-out skin in order to ensure the diagnostic dream, a usage found widely elsewhere (Hamilton 1906: 84-5). It appears that the skin was thought to have a particular power both to consecrate and to open the dreamer to divine influence. There were associated dietary injunctions too, wine and, oddly, beans being forbidden because of their inhibitory effect on dreaming (85).

Nothing that looks very much like modern machinery is in evidence here, but we can be sure that the work done by the ritual is the magico-mechanical kind that is common in almost all therapeutic practices, and especially those with no physical basis. There must always be some procedure to be worked through obediently. This encompasses both the production and the interpretation of the dream. Cure requires the operations of an oneirotechnic that is both actual and imaginary – actual in its operations, and perhaps also in its effects, though imaginary in terms of the mechanism that is supposed to be at work (and in the work of that supposing).

We may suspect that much of the placebo effect, which, given the extraordinary variability in drug efficacy, may operate to a much larger degree in organic medicine than we may imagine, depends upon something of this materialist logic. Always, it appears, there must be the mediation, if not of a material object – classically, some kind of pill (red sugar pills are routinely found to more effective than blue ones, except in Italy, where the national football team are the 'azzurri' and so perhaps blue has the potency commonly attributed elsewhere to red), then of some usually complex medico-technic mediation that approximates to an object, by hardening action into iterative object-form. There are things we call 'comfort objects' because objects comfort. Comfort and comfiness may seem soft and eiderdowny, but the word originally suggests that which fortifies or confirms. The verse from Isaiah 41 which the King James Version renders as 'he fastened it with nails, that it should not be moved' is given in the Wycliffite Bible as 'He coumfortide hym with nailes that it shulde not be moued'. This is how the rod and the staff of Psalm 23 can be said to 'comfort', a sentiment otherwise intelligible only to sexual enthusiasts of a specialised kind.

Unlike the Gysin Dreamachine, which was designed to induce dreams, the dream machine devised by sleep researcher Keith Hearne during the 1980s was designed to facilitate investigative control over the dreaming process. Hearne's dream machine was in fact nothing more elaborate than a respiratory monitor which measured changes in rates of breathing that can be correlated with the periods of REM sleep in which dreaming occurs. However, Hearne also discovered that being able to detect automatically when dreaming was occurring in a sleeping subject made it possible to direct the dream-process in various ways, for example by introducing a physical stimulus that might be incorporated into the dream narrative. It even proved possible with certain subjects to trigger the state known as 'lucid dreaming', rousing a dreamer by a coded series of electrical pulses sufficiently for them to be able to observe, direct, and even to report on their dreaming in 'real time'. Hearne is clear that the purpose of the dream machine is not that 'before you sleep you somehow "programme in" the dream you wish to have' (Hearne 1990: 97), though this does seem implicitly to be promised in the very idea of a technology that allows for conscious control of dreaming.

One of the interesting features of the dream-detecting apparatus was that it tended itself to become assimilated into the dream-content, as for example in one dream reported by a subject, which Hearne suggests may be a representation of the dream-machine itself:

> I was walking into a house. Music was coming from a chest of drawers. I 'knew' that each compartment or drawer can be changed to another section so causing the sound to change. I am trying to decide how to change the volume, when I am woken. (Hearne 1990: 36)

It is not clear whether in a lucid dream the dreaming subject is to be regarded as being awake and aware while dreaming, or as dreaming that they are awake. After all, dreams of being awake, or of lying awake unable to go to sleep, are a common occurrence, so there seems no reason in principle why one should not also be able to dream that one is lucidly dreaming. The mediation of a dream machinery seems to assist this process of transforming the dream from an experience to a manipulable kind of event.

If the dream can be thought of as mechanical, it can also be generative of mechanism. Just as scientists and inventors have often reported solving technical or theoretical problems in dreams, so it has recently been suggested that one might, so to speak, mechanically harness dream capacity for engineering purposes. Deirdre Barrett has reported on techniques for controlling dreams in order to make them instrumental, Dreaming may be cognitively useful in this respect because it intensifies spatiovisual awareness (alpha rhythms being associated with the production of visual imagery) and perhaps also 'mutes' language function. Barrett recommends priming the dreamwork in a series of 'incubation instructions', which include the following steps: writing the problem down as a brief phrase or sentence, and placing it by your bed; visualising the difficulty as a concrete problem, and visualising yourself as successfully dreaming the problem's solution; arranging objects connected with the problem on your night table; lying quietly on awakening and writing down any dream memories (Barrett 2001: 120). Another way of describing the process of rendering a problem in spatiovisual terms is to see it as a mechanisation of the problem. Indeed, one might almost say that to turn a kind of intellectual difficulty into a problem is itself a process of devising a kind of mechanism capable, as we say, of 'working out' a solution. I have myself sometimes posed a question to myself in this way before going to sleep and at least had the sensation of having worked through to some kind of answer to the problem on awaking. It may however be that the largest part of the work involved had been done in simply framing the question in the first place; good teachers know the value of helping a student to reconfigure an intellectual impasse into a problem capable of being analysed into a series of moving parts and yielding a definite result.

It is interesting that the procedure may include the act of dreaming itself, making of the incubation procedure a kind of reflexive design technology: the dream that imagines its own dreaming process as a kind of machine in order to facilitate the dreaming of a more perfect machine. This process seems to be assisted by the move from language to visual or motor forms. Seemingly, many people find they cannot easily read text in a dream – the letters they see and recognise as writing may often be illegible, or mutable. But of course writing is not necessarily opposed to spatio-motor form, for one might see writing as in certain respects a mechanisation of speech. This might suggest that writing itself can be thought of as a kind of dream machine. In a remarkable nineteenth-century text on dream production, Hervey de

Saint-Denys suggested that the mode of analysing dreams as allegories, as influentially embodied in the work of the second-century Artemidorus, is a consequence of the widespread belief among the Egyptians that the gods who sent dreams were also the originators of writing systems: 'nothing more natural, then, than to suppose that the same gods whom they took to be the authors of dreams, employed the same hieroglyphic language' (Hervey de Saint-Denys 1867: 54n; my translation).

The mechanism for dream-incubation and recall suggested by Barrett involves written rather than spoken mediation: writing out a problem, keeping a torch and pen by the bed, writing down the solution, rather than, say, reciting the dream out loud. Charles Dodgson went further and actually devised a machine for writing at night without the need to get out of bed. Dodgson's 'nyctograph' consisted of squares in which one could write one character at a time. Dodgson improved on the device by inventing an alphabet of dots and lines adapted to the squares (Douglas-Fairhurst 2015: 316). A solution so far in excess of the problem it is meant to solve suggests that the dreamwork has overtaken its purpose.

So one mechanises the dream to make it an instrument of machine-production. August Kekulé's famous dream of the structure of the benzene molecule (Read 1995: 179-80) may be of this kind, along with Mendeleyev's dream of the periodic table (Strathern 2001: 282-6), for both of these are schematic structures capable of being thought of as mechanisms. William Blake described being told by his dead brother Robert in a dream about a method for doing hand lettering in reverse, which was a crucial part of the process of engraving employed in his *Songs of Innocence* and other works (Erdman 1977: 100). Elias Howe, the inventor of the sewing machine, described a dream in which he was commanded on pain of death to complete his design for the machine, which he had until that point not been able to make work, with a hole in the middle of the needle shank. In one account of his dream

> he saw himself surrounded by dark-skinned and painted warriors, who formed a hollow square about him and led him to the place of execution. Suddenly he noticed that near the heads of the spears which his guards carried, there were eye-shaped holes. He had solved the secret! What he needed was a needle with an eye near the point! (Harrington 1924: 385)

Another dream-mechanism was devised, or at least reported, by D.B. Parkinson, a researcher at Bell Telephone Labs. Parkinson had devised a potentiometer, a device for recording fluctuations in voltage. In 1975, he recorded having the following dream in the spring of 1940 as German forces swept across Northern Europe:

> I had been working on the level recorder for several weeks when one night I had the most vivid and peculiar dream. I found myself in a gun pit or revetment with an anti-aircraft gun crew [...] There was [a] gun there [...] it was firing occasionally, and the impressive thing was that every shot brought down an airplane! After three or four shots one of the men in the crew smiled at me and beckoned me to come closer to the gun. When I drew near he pointed to the exposed end of the left trunnion. Mounted there was the control potentiometer of my level recorder! There was no mistaking it – it was the identical item. It didn't take long to make the necessary translation – if the potentiometer could control the high-speed motion of a recording pen with great accuracy, why couldn't a suitably engineered device do the same thing for an anti-aircraft gun? (Mindell 1995: 73)

In 1961, John Whitney adapted the high-speed position-plotting apparatus used in Second World War gunsights to produce *Catalog*, a film of computer-generated visual effects. Ian McFadyen connects this with the story that the inventor of the sighting device had in fact 'seen the robotic mechanism in a dream and had drawn it when he awoke', implying then that Whitney had turned the dream-produced device into a device for producing dreams (Cecil 1996: 23). As I have just noted though, Parkinson did not record his dream until more than decade after the making of the film – did Whitney anyway somehow see, or hear of it? In any case, it is a writing-act – the inscription of the dream of a machine that translates one kind of writing into another – that here provides the mediation between the dream and the machine.

Perhaps we dream through, with, and about objects ultimately because the dream borrows or, in Kleinian terms, introjects, certain kinds of object-form, or substitutive substance, to keep itself in being. The form that Bertram Lewin proposed for this in 1946 was what he called the 'dream screen', an imaginary surface which represents the satiety and containment of the breast, an imaginary integument which maintains the integrity and intactness of the dream (Lewin 1946). If the purpose of the dream is in part to keep the sleeper asleep, by soaking up distractions and disturbances and digesting them into narrative, making the dream a machine for converting noise into information, then such a temporal continuity-function might aptly be embodied in a continuous object like an endless unrolling film.

Lewin's quasi-cinema is anticipated by Hervey de Saint-Denys in his detailed investigation of dream-processes and recommendations for directing them. Hervey de Saint-Denys explains that dreamers seem to have the capacity to dream of complex visual forms like buildings without having any architectural or engineering training because dreams work with photographic '*cliché-souvenirs*', images that have previously 'photographed themselves' and been stored in the memory, unknown to the subject until they return in dream, in a 'mysterious process which works spontaneously' (Hervey de Saint-Denys 1867: 32; my translation). So here the imaginary camera of memory is supplemented by the magic lantern, which is Hervey de Saint-Denys's favoured metaphorical apparatus, allowing as it does for forms of overlayering, or double exposure (33, 40-2), in what Freud would call 'hypermnesic' dreams (Freud 1953-74: 4.13). Though he draws on ideas of mechanism to explicate the dreaming process, Hervey de Saint-Denys was determined to assert the power of the will in and over dreams. He argued against purely physicalist interpretations of dreams as the product of morbid stimulation of the nerves (Hervey de Saint-Denys 1867: 161) and against the materialist 'mania' (74) of a commentator such as Boerhaave, who maintained that dreaming is a state of delirium, in which, in a passage quoted by Hervey de Saint-Denys, 'one has no more than a mechanical existence', 'on n'existe plus que machinalement' (75). Hervey de Saint-Denys resembles other writers on dream machinery in his tendency to imagine positive and negative modalities of the machine, depending on whether the dreamer is the dream's producer or production.

There is a similar relation between the dream-machines that one makes one's own through a kind of active engineering and the invasive and (usually) oppressive mechanisms that Victor Tausk in 1919 described as 'influencing machines'. These machines, the most famous subject and exponent of which was Daniel Paul Schreber, are imagined to control the thoughts and sometimes also bodily sensations of their subjects from a distance, typically through waves, rays, or wires. In one sense, they are tyrannous and persecuting; but they also offer a kind of pleasure in the possibility of a kind of surrogate remote control of the mechanism through the detailed explications, either in verbal or in visual form, that sufferers from such systematic delusions (delusions of systems and systems of delusion) generate. The subject is driven by the machine that works upon him or her to a demanding and often deeply absorbing work of self-production, an inscribing of delusion that is far from being itself deluded (Connor 2010: 43-101).

Tausk agreed with Freud's suggestion in *The Interpretation of Dreams* that 'all complicated machinery and apparatus occurring in dreams stand for the genitals, and as a rule male ones – in describing which dream-symbolism is as indefatigable as the "joke-work"' (Freud 1953-74: 4.355). But in Freud's own work, machines tend to have a rather different signification. They tend in fact to symbolise psychoanalytic treatment, or psychoanalysis itself, the very work that is being done to reveal the work of the dream. Freud relates the dream of a female patient: 'She was in a big room in which all sorts of machines were standing, like what she imagined an orthopaedic institute to be. She was told I had no time and that she must have her treatment at the same time as five others. She refused, however, and would not lie down in the bed – or whatever it was – that was meant for her' (199). Freud writes that

> [t]he first part of the content of this dream related to the treatment and was a transference on to me. The second part contained an allusion to a scene in childhood. The two parts were linked together by the mention of the bed. The *orthopaedic institute* referred back to a remark I had made in which I had compared the treatment, alike in its length and in its nature, to an *orthopaedic* one. (200)

This hall of machines recurs in one of Freud's own dreams, in which once again the machinery seems to image the apparatus of psychoanalytic interpretation, on this occasion complicated by the fact that the dreaming Freud is assailed by the sense that he is being accused of dishonesty:

> conscious of my innocence and of the fact that I held the position of a consultant in the establishment, I accompanied the servant quietly. At the door we were met by another servant, who said, pointing to me: 'Why have you brought him? He's a respectable person.' I then went, unattended, into a large hall, with machines standing in it, which reminded me of an Inferno with its hellish instruments of punishment. Stretched out on one apparatus I saw one of my colleagues, who had every reason to take some notice of me; but he paid no attention. I was then told I could go. But I could not find my hat and could not go after all. (336)

Here, the machine is perhaps emblematic of the 'dream-work' itself, of the dream as a kind of encoding and partially decoding machinery.

It is scarcely surprising that a work like *The Interpretation of Dreams* that spends so much time exploring and explicating the operations of what Freud calls 'the complicated mechanism of the mental apparatus' (75) involved in producing, remembering, forgetting, and interpreting dreams, should sometimes find this apparatus taking on objective form in the dreams that it subjects to analysis – one wonders whether Freud's dreamers might have been primed by Freud's own mechanical lexis in describing dreamwork. But these dreamed machines also seem like a kind of mocking travesty of the principle that Freud carefully articulates, that this psychic apparatus is to be thought of as logical rather than locative. For, like Babbage's Analytical Engine or Turing's Universal Machine, it consists of relations and so can be made of anything:

> we shall be obliged to set up a number of fresh hypotheses which touch tentatively upon the structure of the apparatus of the mind and upon the play of forces operating in it. We must be careful, however, not to pursue these hypotheses too far beyond their first logical links, or their value will be lost in uncertainties. Even if we make no false inferences and take all the logical possibilities

into account, the probable incompleteness of our premises threatens to bring our calculation to a complete miscarriage. No conclusions upon the construction and working methods of the mental instrument can be arrived at or at least fully proved from even the most painstaking investigation of dreams or of any other mental function taken in isolation. (511)

Like Hervey de Saint-Denys before him, Freud then turns to the example of actual machines, somewhat paradoxically to explain the ways in which the psychic apparatus is not to be reduced to any simple physical arrangement:

> I shall carefully avoid the temptation to determine psychical locality in any anatomical fashion. I shall remain upon psychological ground, and I propose simply to follow the suggestion that we should picture the instrument which carries out our mental functions as resembling a compound microscope or a photographic apparatus, or something of the kind. On that basis, psychical locality will correspond to a point inside the apparatus at which one of the preliminary stages of an image comes into being. In the microscope and telescope, as we know, these occur in part at ideal points, regions in which no tangible component of the apparatus is situated. (536)

Perhaps we should say that what characterises dreams and dreaming is precisely a kind of essential *manque-à-être*, a failure to be, or be anything (any kind of thing), in themselves. The imaginary work of producing dreams may then be continuous with the dreamwork – the work of dreaming and the dream of working – of which they undecomposably consist. Freud seems to have come to recognise this, adding a footnote to the *Interpretation of Dreams* in 1925 in which he corrects not only lay dreamers who mistake the dream for its manifest content, but also dream-analysts who mistake the dream for its latent, or encoded content: 'At bottom, dreams are nothing other than a particular *form* of thinking, made possible by the conditions of the state of sleep. It is the *dream-work* which creates that form, and it alone is the essence of dreaming – the explanation of its peculiar nature' (506 n2).

Dreams are the work they perform on themselves in order to form themselves as coherent or substantial forms. This means that they must include the work of fabulation that Freud called 'secondary revision' (487-507) – dreams are already secondary revision in the never-apparent 'first place'. We may compare this machinery of self-production with Conrad's vision, in a letter to Cunningham Grahame of 1897, of the 'knitting machine' of the cosmos:

> There is a – let us say – a machine. It evolved itself (I am severely scientific) out of a chaos of scraps of iron and behold! It knits. I

> am horrified at the horrible work and stand appalled [...] And the most withering thought is that the infamous thing has made itself; made itself without thought, without conscience, without foresight, without eyes, without heart [...] It knits us in and it knits us out. (Conrad 1983: 425)

The rapture reported by those who are able to supervise and control their dreams in lucid dreaming (or those at least who dream of or with this kind of rapture) perhaps has something to do with their success at producing the dream as controllable object, or ideal psychic mechanism. By contrast, the horror or nightmare reported by Conrad is perhaps that of the dreamer lucid enough only to find himself part of the machinery of self-production and self-entanglement without there being any position from which the dream-machine may be operated. Perhaps part of the rapture of the lucid dream, continuous as it may be with Freud's suggestion that art may be a sort of command-and-control daydreaming, is the overcoming – or, as we should perhaps rather say, the dreaming away, or phantasmal overcoming – of the resistance to control that seems to be part of many dreams, as embodied in the fact that dream mechanisms or structures are so often unreliable, perverse, or paradoxical. One of the most striking indications of this is what has been called the 'light-switch' phenomenon. Lucid dreamers were asked to imagine turning on a light switch in their dreams. Almost all reported that the switch did not work – the lights failed to come on, came on in the wrong place, or produced a dysfunctionally sparking and flickering light-bulb rather than full illumination (Hearne 1981: 98). One of Hearne's subjects reported thinking that it was 'typical of this place, nothing works properly' (98), which assuredly applies to most of the gadgetry in my dreams. Hearne suggests that, because the lights suddenly going on is associated with the interruption of the dream in waking (as well as, we might add, the end of a film or play), this may be the dream's self-defence, or the sleep-maintaining function that dreams may perform, pointing to the existence of what Hearne calls an 'autonomous dream-producing process' (98). The machine of dreaming that is designed to ensure that the dream continues at all costs disables all other mechanical operations that might override its workings. The default machinery of dream engineers machine-deterrence. Even for the so-called lucid dreamer, there can be no full *fiat lux*.

Machines are necessary mediations between the things called things and the no-things called selves. I am and am not a machine in the same way as a machine is and is not an object. I am a machine's self-exceeding as a machine is an object's self-exceeding. A machine is an object that acts, as well as is. A mind is a machine that feels itself acting, or feels, or wills the feeling, that it does. But the machine can never be entirely left behind in this self-exceeding, precisely because it is the machine and not some other thing that is exceeded, and also because the exceeding is anyway part of how the machine works.

Where I feel, or dream myself to be in relation to this exceeding of machinery programmes much of the feeling-tone – whether fascination, dread, rage, or delight – invested in dream machines or generated by the machinery of dream.

The idiom I have proposed for reading such objects, or objectifying apparatuses, is a psychotechnography – even though one would find oneself in short order saying that that psychotechnography is exactly what dream machines already are. The term itself implies a kind of *combinatoire* for conjugating its three components, of *psyche*, *techne*, and *graphesis*: the writing of dream machines, the writing of the machinery of dreams, dreams of writing machines, dreaming of the machinery of writing, machines for writing dreams and machines for dreaming of writing. That's it: as bell-ringers and players of fantasy-tournaments will know, 3! (3 factorial), or 6 triads gives the complete set: WDM; WMD; DWM; DMW; MWD; MDW. But the *psycho-* prefix signifies more than just imagination or fantasy – or signifies that force that makes fantasy more than just fiction or falsehood. Fantasy is the forceful existing, or will to being, not only of what does not exist but also of what does. Fantasy is always both the force of feeling and the feeling of force, a field of forces and feelings that is ideally mediated by dream machines. It is the insistence on existence that makes for objects by injecting the *it must be* into the *it is*. It is the impassioning of the object that substantiates itself, writes itself up as subject substituting itself in the dream machine.

Let me try to reduce what I have been saying to its elementary working parts.

- Machines are what we dream of, and what we dream with. Whether or not they are of machines, dreams seem machine-like.

- Emotion is always meshed with the motor force of these machineries. Dream machines machine dream feeling. And none of the wanting, fearing, mourning, envy, lust, disgust, horror, or fascination can occur without the mediating apparatus of the dream machine. So the feelings we have about machines are feedback mechanisms, in that a proportion of the force of whatever we may feel about machines must be borrowed from machines themselves.

- Not all objects are machines; but all machines are dream machines.

4

Pleasure Machines

Pleasure machines bring to the fore the question of the relation between feeling and machines as such introduced in Chapter One. We always feel for machines, we have feelings for as well as about them. Our feelings for machines are provoked in part precisely because machines are not supposed to have feelings, either for us, or for themselves. To invest anything with feeling, with negative or positive value, is to draw it into the phantasmacopia of attachments and revulsions that is human life, the great drugged, rugged dream of our relation to the world. Feelings about things are always a form of judgement on them, to be sure, but feeling is always as well the flooding of the world with fantasy, and never more throbbingly and exorbitantly than when it is feeling for the way things are *an sich*, in or toward themselves, the selves we know they cannot have. So if there are no machines that do not arouse affect, there is no affect and in particular no machine-affect, without fantasy. This is another sense in which all machines about which, or through which, we feel, are imaginary machines, and the ways in which we feel about them a phantasmotechnics.

In Brittle Orbs My Labours Acted

The deep and complex involvement of human beings with machinery is signalled by the degree of the pleasure that we take from operating machines. There may be pleasure, for example, in the awareness of how smoothly and skilfully one engages with the machine, understanding its limits and possibilities and making it work reliably and responsively to our needs. Then there may be a different kind of pleasure involved in getting the machine to do something unexpected or that it was not designed to do. But perhaps the pleasure of engaging with machines largely reduces to the subtractive pleasure gained simply from not having to do the work that the machine is doing for you. We count on the pleasure we derive from the calculation of what for the first time in 1791 were called 'labour-saving' devices. But in order for there to be pleasure in the thought of labour saved, there needs to be some accessory work done, in the formation of that thought, which will need to take the form of some kind of calculation, or at least estimation of the difference between actual and potential labour. (The pleasure principle will perhaps always foil the operations of Maxwell's demon.) Only in representation can there be minus quantities, and only in the kind of imaginary machine we call calculation is the representation possible of a minus quantity as a positive. The pleasure in fact comes, not immediately from the machine that is doing work instead of you, but from the mediating operation of this machinery of calculative representation. When I reflect pleasurably that my lawnmower, electric razor, washing machine, or trouser press is doing the work that I might otherwise and more wearisomely have to do, my pleasure depends upon this subsidiary piece of work that I am doing of subtractive calculation.

In his *Mathematical Magick* (1648), John Wilkins set out to show some of what his subtitle called '*The Wonders that May Be Performed by Mechanicall Geometry*'. The explication of the principles of the balance, the lever, the cog-wheel, and the screw leads Wilkins to a mathematical demonstration that it is possible 'for any man to lift up the greatest Oak by the roots with a straw, to pull it up with a hair, or to blow it up with his breath' (Wilkins 1648: 96); along with, even more remarkably, a vindication of the Archimedean claim that, given a lever of the right length, it would be possible to lift the weight of the whole world (101). Wilkins makes it clear that these extraordinary scalings-up of weight and force must always be paid for in speed – so that though it is possible to lift a hundred-pound weight with the force of one pound, it will take a hundred times as long. Wilkins represents this limit as a curb on vaulting imagination:

> If it were possible to contrive such an invention, whereby any conceivable weight may be moved by any conceivable power, both with the same quicknesse and speed (as it is in those things which are immediately stirred by the hand, without the help of any other instrument) the works of nature would be then too much subjected to the power of art: and men might be thereby incouraged (with the builders of *Babell*, or the rebell Gyants) to such bold designes as would not become a created being. And therefore the wisdom of providence hath so confined these humane arts, that what any invention hath in the *strength* of its motion, is abated in the *slowness* of it: and what it hath in the extraordinary *quickness* of its motion, must be allowed for in the great *strength* that is required unto it. (104-5)

And yet Wilkins is not thereby inclined to abate his own speculations in mechanical geometry, taking the latter word literally as he moves to a cosmic scale to show 'that 'tis possible Geometrically to contrive such an artificiall motion, as shall be of greater swiftnesse, then the supposed revolutions of the heavens' (144). His caution to his reader that 'the quaere is not to be understood of any reall and experimentall, but only notionall, and Geometrical contrivance' (142), both curtails and releases his speculative mathematics, the mechanism of the calculation at once substituting for and seeming to promise

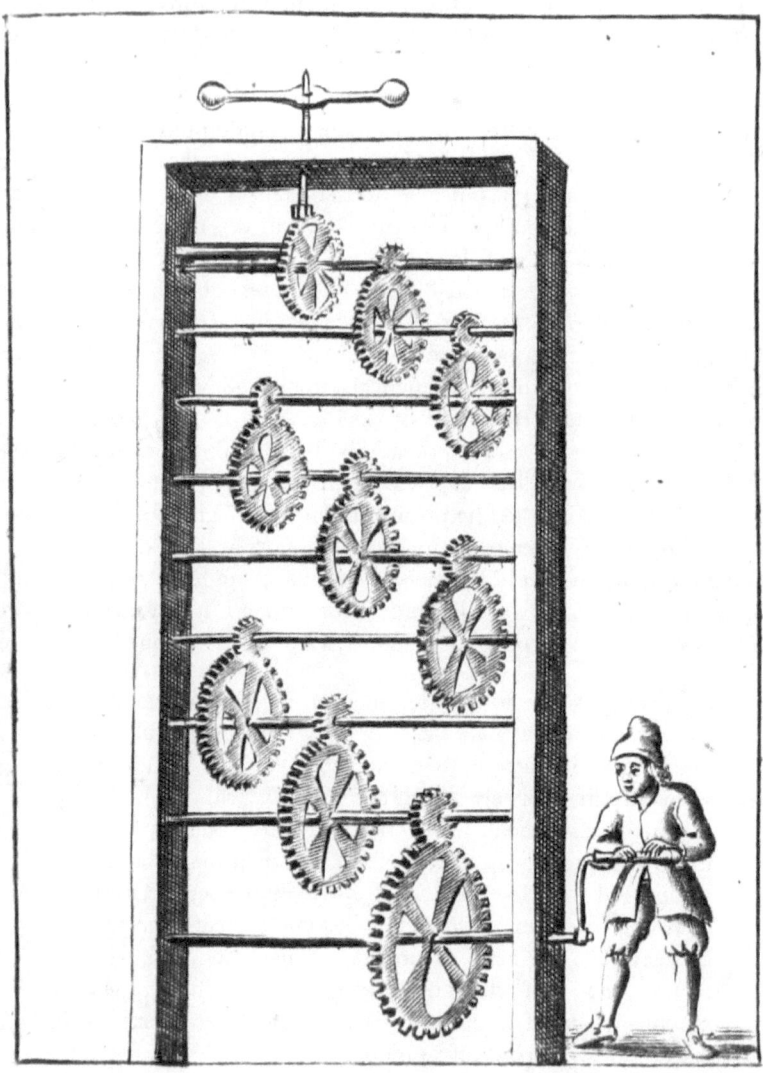

Fig 1: Wilkins 1708: 83.

the possibility of actual application. The illustration he provides is both a demonstration of how quickly mechanical gearing can multiply speed and an imaging of the proliferative powers of imaging itself (Figure 1).

This kind of calculative representation is an imitation and iterative echo of the work involved in conceiving and designing such a machine in the first place. It is another of the many modes in which machinery is written, or constituted in commentary and picturing. It also indicates how representation and affect are geared together; for to represent the workings of a machine is to show the mechanical imagination at work. The pleasure of the model or mechanical scheme derives, we might say, from the disproportion between the huge outlay of energy required to produce large mechanical effects – prodigious speed, the lifting of huge weights – and the relative ease with which they may be produced mechanically. There is always a homology between the economy effected by the schema with respect to the actual construction and operation of the machine and the economy effected by the machine with respect to the work to be performed in the world. The schema itself performs a kind of work, in anticipating the way in which the work itself turns the world into schema or reveals its schematic basis. These different levels are locked together. A machine is always an image or, in that telling phrase, a 'working model', a model of a certain kind of work which itself performs the work of modelling, of the hidden form of the world.

No machines seem to bring together the small and the large, blueprint and working model, as satisfyingly as the mechanical models of the heavens produced in the ancient world. Wilkins refers to one of the most famous of these, the spherical model made by Archimedes as described in the *Sphaeropoiea*, the book he wrote describing its workings. Neither the object nor the book survives, but the memory of the contrivance prompted a poetic celebration by the Latin poet Claudian 600 years later. Wilkins quotes it in *Mathematical Magick*, in a rather good contemporary translation by Thomas Randolph:

> *Iove* saw the heavens fram'd in a little *glasse*
> And laughing, to the gods these words did passe;
> Comes then the power of mortall cares so far?
> In brittle orbs my labours acted are [*Iam mens in fragile
>   luditur orbe labor*] [...]
> And viewing now her world, bold industry
> Grows proud, to know the heavens his subjects be,
> Beleeve, *Salmoneus* hath false thunders thrown,
> For a poor hand is natures rivall grown (165)

Wilkins thinks that it is most implausible 'that this Engine should be made of glasse', and surmises instead that '[i]t may be the outside or case was glasse, and the frame it self of brasse' (165). Glass is of course better equipped for showing than for working: but if the point of such a machine is precisely to show its own workings, then the fragility of the orbs may be functional, its labour being literally the way in which mind is put into play – *in fragile luditur orbe labor*. Wilkins is known principally as one of the moving spirits of the Royal Society and for his scientific speculations on language, mechanics, and communication. But he was also Bishop of Chester from 1668 to his death in 1672, and as such was as concerned with the mechanics of rhetoric as with

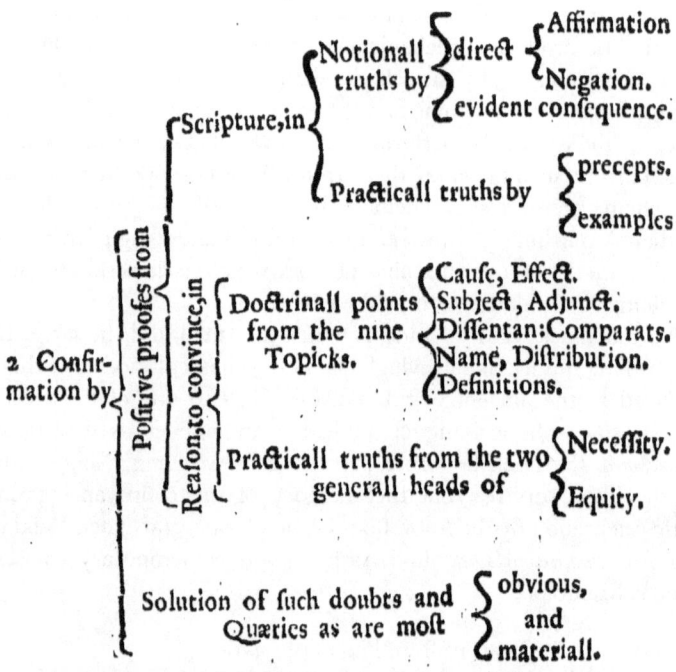

Fig 2: Wilkins 1646: 6.

the art of mechanics. His *Ecclesiastes, or a Discourse Concerning the Gift of Preaching* (1651) describes preaching as

> such an expertnesse and facility in the *right handling and dividing the Word of Truth, as may approve us to be Workmen that need not to be ashamed.*
>
> It does require both { Spiritual / Artificial } abilities. (Wilkins 1651: 4)

Wilkins even lays out the components of the arts of preaching in a kind of schematic machinery (Figures 2 and 3).

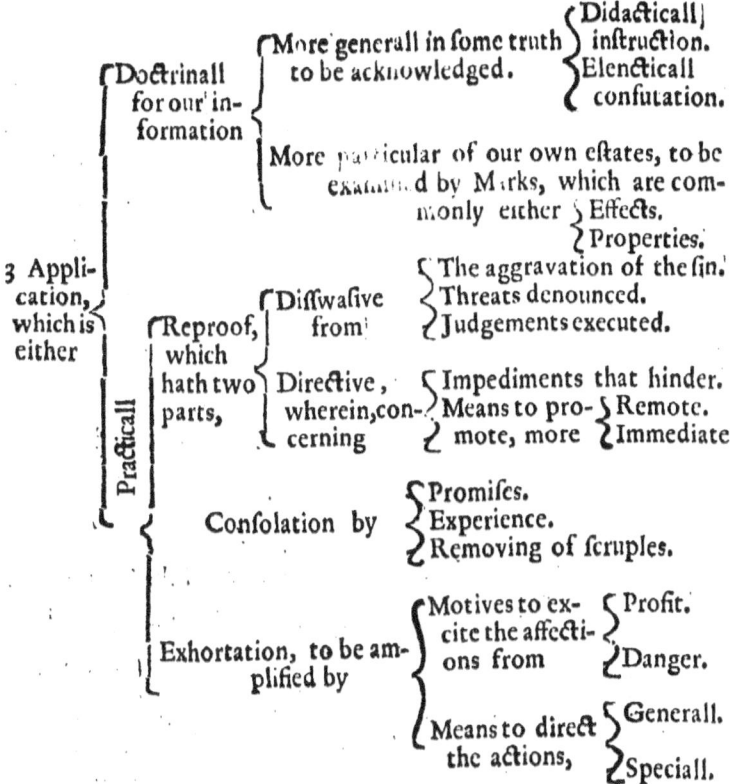

Fig 3: Wilkins 1646: 7.

Comforts

I want in this chapter to show that pleasure-machines always conjure and conjoin with imaginary machineries of pleasure. Since the quasi-mechanical anticipation and recapitulation of pleasure seems to be such an important part of the relations that humans establish with machines, it should not be a surprise to find that humans conceive machines the primary purpose of which is to give pleasure, rather than producing it as an accidental by-product. During a very hot summer in 1990, we set up an electric fan in the bedroom of our one-year-old son to help him sleep. Injudiciously, we left it on the oscillating setting, which meant that, when we went to check on him an hour or so later, we found him not asleep, but pulling himself energetically but absorbedly to left and right by the rails of his cot, to keep the cooling flow playing on his face and hair. The work he was performing in pursuit of his pleasure was undoubtedly heating him more than the fan was cooling him, but he had made an estimation of pleasure to be gained – or rather, perhaps, he had made himself over, or been himself made over by his pursuit of pleasure, into a machinery of pleasure-estimation.

All pleasure machines involve an active-passive calculus in the relation of the self to itself. The machine of pleasure is not only one modality among others of Sartre's for-itself, it is perhaps the machine-relation that is essential to this reflexivity. The machine substitutes, as I have said, performing the work that I might myself otherwise have to do. But pleasure machines work differently from ploughs or steam-irons, in that we use them to work reflexively on ourselves. Indeed, machines of pleasure typically participate in one of the most striking physical features of the human body, namely that it can reach almost every part of itself, performing for itself actions which in other species requires cooperation from another member of the same species, or of another. My cat can reach with her tongue or claws almost every part of herself but the top of her head and her chest, and so must enslave me with an imitation of helpless cuteness, and a rather impressive little pantomime of scratching with her left paw circling in air, in order to get me to act as her frica-wallah. We may take frication as a primary form of self-pleasuring and, in its association with grooming, among the most important ways in which relations of pleasure-giving are formed in, and thereby themselves form, primate groups.

When I scratch myself, I divide myself into two, the scratcher and the scratched. More precisely, perhaps, I exploit the division of my skin into two forms, the soft, innervated, and sensitive flesh, and the hard, horny, insensate keratin of my hair and fingernails. This keratin is related to the claws, hooves, feathers, shells, scales, beaks, and horns of other animals. It is keratin that joins the cornea in the human eye to the horn of the rhinoceros. It is hard to think of any exercise of subtlety and dexterity that does not require some kind of mingling or cooperation of the soft and the hard components of my body. I can only have a self-relation if I can divide myself to form a hard-soft coupling, in which the hard can be brought to bear on the soft. If I am to play a stringed instrument with any sensitivity and precision, I will need to reduce the sensitivity of my fingertips, by building up the layer of keratinous tissue we know as a callus. In this, the production of music is an image of the hard-soft mingling that is involved in all musical structures. I now find it very hard to sing the songs that have most moved me all my life, because they melt me so infallibly. I have to distance and indurate my voice as I sing, putting it away and aside from me, in order that I can be fully penetrated by it. The only way to know myself fully, or meet fully with myself is to become partially other to myself, which means either literally or metaphorically to incorporate some hardness or numbness. Perhaps all of these routines are summed up in the method that Billy Connolly discovered as a boy for the most effective masturbation, which is to sleep on your hand until you lose all sensation, so that it feels like somebody else is doing it. 'Somebody else' is a much more precise and expressive formulation than the pompous 'Alterity' or 'the Other'.

Masturbation is the bringing to bear on my body of some body-other, some portion of the 'other-body' that my body is in fact so well-equipped to be to itself, inasmuch as it has hard and alien as well as soft and amicable aspects. I am well equipped to appreciate the paradox of the word 'comfort', which, as I noted in Chapter Three, essentially means to strengthen or harden, as I went to a school where we slept on thin mattresses over wooden planks. This fakir-like early formation means that I can sleep soundly on the floor of an airport, and spring up in the morning like a lambkin, while a night spent drowning in the mush of a soft mattress means I awaken stiff as a board.

The reciprocation of the hard and the soft is strongly in evidence in play and gaming. When I play, I must have some object, something hard against which, obedient to its etymology, the softness that I feel myself to be – soft in the sense of being open and full of indefinite possibility – can come up against. All play requires hard objects, which are required for me to know and feel myself in play, which is actually the process of softening the hard, to discover by reciprocation that I am in fact the player. Ted Hughes seems to dramatise this in his poem about a Caliban-like creature unknown to itself using its explorations of the world to give itself shape: 'if I go / to the end on this way past these trees and past these trees / till I get tired that's touching one wall of me' (Hughes 2003: 183). Herb Brody, the editor of *Technology Review*, wrote presciently in the early days of personal computing of the importance of the pleasure relation to computers, quoting examples of users' sense of rapture, which suggested 'reveries on sensual experiences such as sex and chemical intoxication' (Brody 1992: 32). A graphic designer writes that graphics software gives him 'a weightless feeling, like my mind has been freed of the constraints of gravity', while the editor of another technology magazine evokes the 'pure, fugue-state, mind play involved in connecting with the PC' (32). This is no simple dissolution, but rather a ludic relation of hard and soft, the coming and going of effort and resistance. As with a game, there must be a dynamic balance of the user's skill and the challenges of the game. With too much resistance, there is anxiety and frustration. If the game yields too easily, boredom results (34). From the beginning, the PC is designed, and employed, as a pleasure machine.

The idea of a 'pleasure machine' must always be double. It will always mean a machine that gives pleasure, while simultaneously indicating the mechanisation of pleasure. The pleasure which a machine may give is always in part and therefore in addition the pleasure gained from making a machine of pleasure. As a machine for playing with, a pleasure machine is therefore always also a way of playing with one's own pleasure. The automation of pleasure always gives the pleasure of automation. The machine that produces pleasure produces also the pleasure of pleasure production. Surprisingly, the pleasure of automation comes in part from the fact that it is not in fact simply automatic any more, for it is now a machinery that requires to be governed and maintained. This means that pleasure is gingered by being fragile and captious. Beci Carver writes of the 'absolutist' transaction offered by the slot-machine, which always seems to give exactly and invariably what one's coins will purchase (Carver 2016). But in fact the notorious unreliability and exasperating fussiness of slot machines form an important part of the pleasure that attends the successful exchange of coins for candy, a pleasure that is in proportion to the rage vented on the vending machine that reneges on its part of the bargain.

Desiring Machines

There must always be a machinery of representation that requires and allows one to see one's pleasure as a machine. This machinery must involve some work – even, no especially, the work of imagining the kind of machine that would be necessary to guarantee that one need never work again. One of the most provocative and influential accounts of the relations between pleasure and mechanism is the concept of the 'desiring-machine' invented by Gilles Deleuze and Félix Guattari in their *Anti-Oedipus* of 1972. Now, the most striking feature of the Deleuzian desiring-machine is how entirely unmechanical it is, that is, how little work it requires. This is because what Deleuze and Guattari designate as desiring-machines are in essence any kinds of conjunction which enable some kind of flow (usually of some kind of 'energy'). This flow is always, it seems, just there already and so needs only to be channelled by the desiring-machine. Deleuze and Guattari emphasise that there is always interruption where there is conjunction, but interruption always seems accessory to conjunction too, and works always to amplify and impart greater intensity and extent to it.

Their explication mimics this rhythm of enlargement, with conjunction at each stage overcoming impediment, and seemingly producing itself out of it. There is first of all the 'paranoiac machine', in which machineries of different kinds cramp and confine the 'body without organs' of pure unstructured sensation:

> Every coupling of machines, every production of a machine, every sound of a machine running, becomes unbearable to the body without organs [...] This is the real meaning of the paranoiac machine: the desiring-machines attempt to break into the body without organs, and the body without organs repels them, since it experiences them as an over-all persecution apparatus. (Deleuze and Guattari 2000: 8)

But then this 'repulsion-machine' – a kind of defensive bristle against invasive machinery that itself hardens into a kind of machine – gives way to another kind of machine, an 'attraction-machine' this time, in which the persecuted subject starts to identify itself with the agencies and instruments of its persecution. This produces what, borrowing a word from Daniel Paul Schreber's account of his delusions, Deleuze and Guattari call a 'miraculating machine', which allows 'the establishment of an enchanted recording or inscribing surface that arrogates to itself all the productive forces and all the organs of production' (11-12). The miraculating machine is made up of disjunctive syntheses as well, every interruption and diversion the introduction of a new flow: 'Machines attach themselves to the body without organs as so many points of disjunction, between which an entire network of new syntheses is now woven, marking the surface off into co-ordinates, like a grid' (12).

But there remains a contradiction between the repulsive paranoiac machine and the attractive miraculating machine. Once again Deleuze and Guattari propose in place of the contradiction a kind of conduction, in the form of a third kind of machine, a 'celibate machine' formed from the joining together of the two other machines:

> A genuine consummation is achieved by the new machine, a pleasure that can rightly be called autoerotic, or rather automatic: the nuptial celebration of a new alliance, a new birth, a radiant ecstasy, as though the eroticism of the machine liberated other unlimited forces. (18)

The ideal or absolute condition beneath and beyond all of these conditions is that of a body without organs, with its experience of total flow without restraint in all directions, unchannelled by any organ.

> The body without organs is not God, quite the contrary. But the energy that sweeps through it is divine, when it attracts to itself the entire process of production and serves as its miraculate, enchanted surface, inscribing it in each and every one of its disjunctions. (13)

The flow that overcomes resistance has here become identified with every resistance, which exists only to conduct flow. But there could be no flow at all under such conditions. A body without organs would not meet the conditions for being any kind of body at all, since to be a body there must be a difference of complexity between an inside and an outside, an enclave in which entropy is reduced, and therefore some kind of boundary or threshold between the two regions. Deleuze and Guattari's imaginary machines aim to overcome the conditions of possibility for all machines, but in the process stop being machines at all, since, in overcoming every resistance, they deprive themselves of the possibility of doing any actual work. Deleuze and Guattari remark that '[d]esiring-machines work only when they break down, and by continually breaking down' (8), but they seem to mean by this simply that a machine that diverges from its proper function always thereby invents some other function to perform, like a Tinguely apparatus. They mean, in other words, that it is impossible for such a machine to break down. But then such a machine can never really work either, since working means the possibility of failing to work, or working less well.

Deleuze and Guattari offer a symptomatology of schizoid fantasy. But as with other applications of and to the theory of the 'influencing machine', they put themselves partly under the influence they explicate. It is difficult for them – or would have been if they had tried – to keep the conceptual meta-machinery of their analysis of schizoid fantasies separate from the machineries themselves. This is because influencing machines always tend to become both form and content of the analysis offered of them by their subjects, and therefore always liable to become meta-machines, machines for explaining the process by which mechanical delusions are themselves produced. The interposition of this mediating meta-machinery is what enables self-reporters of systematic delusion like James Tilly Matthews, John Perceval, and Daniel Paul Schreber to move from pathology to self-cure, by turning their autobiographical writing into an analytic engine.

Two years after *Anti-Oedipus*, Robert Nozick exemplified what seems to be a contrasting philosophical use of a desiring machine. In *Anarchy, State, and Utopia* (1974), he imagines the possibility of what he calls an 'experience machine'. This machine is a version of the 'Brain in a Vat' thought experiment, itself a version of René Descartes's Evil Demon hypothesis, all of them resting on the proposition that there might be circumstances in which I might be entirely unable to distinguish real experiences (or more accurately experiences of a real world) from simulated experiences. Nozick is even less specific about the nature of this machine than Deleuze and Guattari:

> Suppose there were an experience machine that would give you any experience you desired. Superduper neuropsychologists could stimulate your brain so that you would think and feel you were writing a great novel, or making a friend, or reading an interesting book. All the time you would be floating in tank, with electrodes attached to your brain. Should you plug into this machine for life, preprogramming your life's experiences? (Nozick 2003: 42)

The point is to show that the arguments of hedonic utilitarians are mistaken because most people would and should not choose to plug into the machine. Nozick does not want us to get too caught up in the details of this machine, and starts waving away difficulties almost straight away. 'Ignore problems such as who will service the machine if everyone plugs in' (43). The machine could avoid the limitations of solipsism because 'we can suppose that business enterprises have researched the lives of many others' (42). Nozick allows us to think about the possibility of what were not yet in 1974 known as software updates – periods in which one could unplug from the machine to allow it to be reprogrammed with all the new kinds of experience that might in the interim have emerged. Nozick neither argues for nor against the possibility of such a machine, but rather shows that, were such a machine available, we would not and should not rationally choose it. For what the thought experiment – which here takes the form of a machine-imagining apparatus – makes clear is that 'we want to *do* certain things and not just have the experience of doing them' (43). Not only this, 'we want to *be* a certain way, to be a certain sort of person. Someone floating in a tank is an indeterminate blob' (43). We want reality, in other words, not just the feeling of it.

However, in the way of such things, Nozick is drawn into tinkering with his machine, as he imagines the ways in which the determined hedonist might try to perfect it, for example by making it a means of self-perfection. For example, 'since the experience machine doesn't meet our desire to *be* a certain way, imagine a transformation machine which transforms us into whatever sort of person we'd like to be (compatible with our staying us)' (44). Or we might overcome the objection that the machine does not meet our need to make a difference in the world with the development of 'the result machine, which produces in the world any result you would produce and injects your vector input into any joint activity' (44). But, having introduced the idea of a perfectible series of machines, Nozick again sets it aside: 'We shall not pursue here the fascinating details of these or other machines' (44). Instead, he roundly articulates the principle that 'what we desire is to live (an active verb) ourselves, in contact with reality. (And this machines cannot do *for* us)' (45).

But in asking us not enquire irrelevantly into the workings of the machine, Nozick is also asking us not to enquire into the rhetorical workings of his explication of it. What Nozick means by a machine is something that offers surrogate rather than actual experiences. But there are some things that we do and are that are unthinkable without such vicariant procedures. Do my contact lenses do my seeing for me, in Nozick's sense? Does my spade do my digging for me? Does my hockey stick play my hockey for me? Should I mistrust the interposing machinery of language, which substitutes representable, which is to say substitutable, experiences for real ones? What if I cannot in fact do anything *for myself*, but always require some kind of accessory machinery which might be suspected of surrogacy, of doing for me the doing I should be doing for myself? What if it is mediation all the way down, back, across and forward?

The reason we would not and should not choose such a machine as Nozick evokes is that it is in fact inconceivable, since, like Deleuze and Guattari's miraculated version, it would not be a machine at all if it did not require any adjustment and could not go wrong. Though all machines work to save or change the experience of time, a machine that could abolish time altogether would cease to be a machine for just that reason. Just as an ideal communicative situation in which there were no possibility of misunderstanding would not allow for anything we could recognise as communication at all (zero noise means zero communication, just like maximal noise), so an ideally efficient machine would not qualify as a machine at all. Machines consist of the coordinated alternation of flows and impediments, in which flow always requires impediment. It might be replied that that it is not necessary for the experience machine to be subject to the same physical conditions as actual machines precisely because it is an imaginary machine, and this is just the sense in which it is imaginary. But we need to insist that imaginary machines must function as imaginary *machines*, and that the machine insists in every imaginary machine.

We imagine that machines do things for us and thereby make it unnecessary for us to do them for ourselves. There certainly are circumstances in which machines do kinds of work that thereby relieve us of the necessity of doing that work for ourselves. But I began by saying that machines always in fact require some work from us, even and especially if it is the work of representation. Nozick's own explication of the machine seems to prove this, since his is a machine that seems to need endless refinement. He starts out trying to show that a machine can never supply experiences for us because machines do not really allow us to overcome obstacles and limits, but he keeps giving glimpses, briskly set aside though they are as not really relevant to the question, that overcoming limits is one of the things that the work of designing and maintaining machines is for. The reason for plugging into the machine is that one would not need to bother about monitoring the machine; but then the real reason for not plugging into the machine is that then one would not be able to change the plug, adjust the voltage, etc. Nozick hopes to distinguish between an entirely passive, machine-mediated existence and an active, unmediated relation to the things that we really are and do. But the things we really are and do are always in fact mediated in some way by some kind of machinery. It is not that machines do things *for* us, or not only that, it is that we do whatever we do *with* them, borrowing our natures from the machines that borrow their capacities from us. Nozick evokes the machine in order to try to stop us thinking about machines altogether, but succeeds only in getting himself and us interested in the possibilities of the machine he evokes.

Nozick does not want his reader to get tangled up in the idea of machines, though it turns out he really himself should be more interested in them than he is. For, in using the idea of a machine to argue against the principle of ethical hedonism, the idea that we might just want pleasurable experiences, he inadvertently discloses the mechanical nature of many if not all experiences of pleasure, which is to say, the principle that pleasure requires effort overcoming resistance. Far from being dispensable in imaginary machines, this is a principle of their functioning. There is a telling moment in Nozick's short explication, which is prompted by the reflection that we would resist being plugged into the experience machine, because it would limit us 'to a man-made reality, to a world no deeper or more important than that which people can construct', whereas we desire '*actual* contact with […] deeper reality' (43). This desire for transcendence

> clarifies the intensity of the conflict over psychoactive drugs, which some view as mere local experience machines, and others view as avenues to a deeper reality; what some view as equivalent to a surrender to the experience machine, others view as following one of the reasons *not* to surrender! (43-4)

This remark seems to acknowledge that there may well be some ways of conceiving an experience machine that take us out of, rather than stranding us within, local human conceptions of reality. This is an admission that the experience-machine might really be a way of doing something, rather than, or in the process of, having something do it for us.

Most of those who have commented on Nozick's experience machine have been obedient to his desire that we do not worry too much about the mechanical details of his imaginary pleasure machine, and simply use it as a means of bringing questions about experience, desire and choice into focus. But the mediation of the argument through the idea of a machine turns out not to be accidental but essential to the understanding of the very questions of pleasure and agency with which he is concerned. Imaginary pleasure machines of the kind evoked both by Deleuze and Guattari and by Nozick indicate something important about the nature of the pleasure taken in machines of all kinds, namely that the pleasure itself involves mechanical principles, of mediated work, and the modulated overcoming of resistance. Not only are machines useful for gaining pleasure, there is no pleasure without some form of accessory machinery of pleasure gained through overcome impediment. Mechanically-mediated pleasure may in fact be the only kind there is.

This is a principle that is repeatedly available for discovery in the most conspicuous area of mechanical pleasure-production, namely sex machines and other artificial aids for sexual pleasure. At every stage it appears that the aim is to reduce the effort, delay, and inconvenience involved in obtaining sexual gratification. We tell ourselves that machinery is a hideous or ludicrous supplement to sexual pleasure, yet sexual pleasure without some form of desire-extending impediment is not easily conceivable. In 1929, Green Peyton Wertenbaker published a story entitled 'The Chamber of Life' which seems to offer an allegory of this, while anticipating Nozick's experience machine. The narrator of the story meets an inventor called Melbourne who has developed a system of mental telepathy that allows him to transport subjects to a world of complete illusion. The narrator is transported to a utopia in which all labour has been rendered unnecessary, allowing the inhabitants to devote themselves to projects of self-cultivation. The narrator falls in love with his guide, a woman called Selda, who explains that their love is impossible:

> 'But I love you,' I cried in amazement. 'And you love me, too. I know you love me.'
>
> 'That means nothing here,' she said. 'It happens sometimes. One has to accept it. Nothing can be done. We live according to the machinery of the world. Everything is known and predetermined.' (Wertenbaker 1929: 638)

The narrator attempts to drown himself and finds himself in the lake in his home town of Richmond. The lesson seems clear: only unattainability ensures desire. The contemporary application is emphasised by the fact that the narrator is in fact an employee at a movie company. We leave him on his way back to work after the dissolution of his dream: 'It wouldn't do to be late at the office, where I, too, was a maker of sometimes cruel dreams' (639).

It is not just sexual fantasy that has been mechanised. Alien objects have been a staple ingredient in human sexual activity, both singleton and plural, for as far back as we may care to investigate. The use of dildos and dolls have put the play of the subject (soft) and the object (hard) at the heart of sexual activity, making the customary complaint that male fantasy 'objectifies' women harder to sustain, as the excitation of the 'it' proves to be ineradicable from the exertions of the 'I' (Connor 2015: 140-1). There may be something grotesque and often ludicrous in the way in which objects may be taken in to sexual play and the sexual player may be taken in by them, but there seems no way to efface the play of hard and soft, object and idea, aggression and surrender, resistance and yielding, from sexuality, unless one were to see sexuality as a corporeal version of the congress of angels one with another. But if there is any fissure or distance in sexual relations, some mediating object may always arise to plug the gap it thereby keeps gaping. There will always be the possibility of an 'it' to mediate the blessed, blissful I-thou relation, a parasitical third that actually makes the being-as-one of the dyad possible.

Sex-enhancing machines, in common with many other kinds of apparatus for intensifying or autonomising pleasure, have often been deprecated as forms of masturbation, on the grounds seemingly that a machine allows one to do to and for oneself what should really be done to or by others. Machines in this sense seem in fact to be doubly masturbatory because they reproduce the autistic relation of self-use, or as it is more usually styled, self-abuse. The machine is a supplement that artificially extends the subject in order paradoxically to clap that subject solipsistically on itself. The pleasure machine seems to fold together two opposite kinds of inauthenticity, the inauthenticity of giving oneself up to some adulterating principle (idolatry) and the inauthenticity of refusing to allow the world to exist as anything but my fantasy (fetishism).

Automaterotic

Industrialisation threw up many forms of apparatus to supplement and intensify sexual pleasure. Where the sexual instrument has still to be worked, or worked on – the pistoning of the dildo, the come-on to the compliant bedknob or accommodating balloon – the sexual machine mingles hard and soft by working on its own. In *The Technology of Orgasm* (1999), Rachel P. Maines has provided a history of mechanical means of procuring female sexual climax. She connects the development of artificial means of stimulating the female genitals through steam, clockwork, and then electromechanical apparatus to a long, and largely concealed history in which physicians had provided relief to women suffering the multifarious symptoms of 'hysteria'. Part of her argument is that mechanisation allowed male doctors to distance themselves from a practice that, though common, had begun to seem distasteful and somewhat demeaning, while continuing to accumulate financial reward from it:

> Massage to orgasm of female patients was a staple of medical practice among some (but certainly not all) Western physicians from the time of Hippocrates until the 1920s, and mechanizing this task significantly increased the number of patients a doctor could treat in a working day. Doctors were a male elite with control of their working lives and instrumentation, and efficiency gains in the medical production of orgasm for payment could increase income. Physicians had both the means and the motivation to mechanize. (Maines 1999, 3)

The reproach to a medical profession looking to maintain its standing and dignity through the use of machinery coexists with a critique of the androcentric view of female sexuality, which was made uneasy by the fact that vibratory technology seems for so many women preferable to traditional coitus. This is supposed to be troubling for those who wished to maintain the supremacy of vaginal rather than clitoral pleasure, Maines finding in many of the representations of the vibrator 'the classic male fear of sexual inadequacy, to which the new technology adds a threat once associated only with industrial artisans: technological obsolescence' (110). *The Technology of Orgasm* charts the movement of vibrator technology away from medicalising control in the early decades of the twentieth century, as electrical devices of all kinds were marketed for home and private use, usually promising vigour and beauty rather than sexual satisfaction directly. One advertisement by the American Vibrator Company in the *Woman's Home Companion* in 1906 hair-raisingly advised purchasers that the 'American Vibrator may be attached to any electric light socket, can be used by yourself in the privacy of dressing room or boudoir, and furnishes every woman with the very essence of perpetual youth' (quoted 103).

Aside from these two, somewhat contradictory ideas, that men found it natural to try to mechanise the production of pleasure in women (presumably because men are themselves kinds of machines), but are also dismayed to find themselves outdone by these pleasure-machines (Maines 1999, 109), Maines has surprisingly little to say about the way in which machinery and sexual pleasure might have commingled from the late nineteenth century onwards. For both male and female, in Maines's account, vibratory technology was merely instrumental, its importance lying simply in what it was possible to do, or not to have to do, rather than with any phantasmal force exercised by the idea of the sexual machine. But to leave out the phantasmal force of machines is to leave out a good deal. Already, during the 1780s, Dr James Graham had effected an extravagant eroticising of the electrical mechanics of his 'Celestial Bed'. He did everything he could to persuade the listeners to the amatory 'lectures' he delivered, or wrote for handsome ladies to deliver, of the analogy between electrical mechanism and sexual process, aiming no doubt to effect a mechano-corporeal induction of the feelings he evoked:

> Every one knows, that one pole of the magnet strongly attracts, and that the other as violently repels. We know too, that the electrical fire in its different states does the same; and that rubbing common iron with a magnet, in certain directions, will communicate magnetism to it. Thus, love on one side has been known to produce love; and as in the course of excitation in making electrical experiments, the jars will now and then discharge themselves accidentally, or involuntarily, as well as by the safer and more natural application of the smooth round knob, or blunt head of the discharging rod; so in courtship, or in amourous dalliance, between the sexes, involuntary *colliquations* will now and then happen: and at best, courtship and dalliance is no other, than charging the battery on the *one* side, and heating and preparing the genial oven of nature on the *other*, for the concoction and maturation of the genial juice. – The analogy holds good even to the *end of the business*; for the natural consummation, or the *final* discharge or delivery of this electrical fluid – of this *balmy benevolence!* – is in all respects, a true electrical operation. (Graham 1782: 64-5)

So confident is Graham of the capacity of his amorous machinery – both the Celestial Bed and the elaborate rhetorical apparatus of its evocation – that he can afford to tease his audience with comic desublimation, concluding his lecture with the assertion that 'Anatomists tell us, and no one can doubt the fact, that the human body is an hydraulic machine, or a mere heap of vessels, tubes, and reservoirs, containing fluids or liquors peculiar to themselves' (82) – along with a *carpe diem* evocation of 'those men and women who once charmed and captivated all who approached them [who] are now little better than old, crazy, neglected machines, whose exhausted reservoirs, and worn out springs, are ready to fall into dust, on the least rude motion' (90).

Carolyn Thomas de la Peña has shown that there were as many forms of electrical supplement for men during the late 1800s and early 1900s as there were for women, though items like the Pulvermacher Electro-Galvanic Belt were not designed to combat the problem of unsatisfied demand so much as the problem of sexual energy depleted by masturbation:

> One might continue to believe that masturbation was a waste of vital energy, but if wearing a belt in the afternoon could replace the supply lost that morning, it would not be a practice one had to avoid [...] When one considers that it was perhaps the psyche as much as the body that needed treating, the true power of electric belts emerges: belts allowed men to envision a body immune to the either-or choice between sexual pleasure and physical health endemic to the Victorian era. Technology suggested one could have both. (Peña 2003: 156)

The technologies of orgasm by no means simply displaced the vaginal in favour of the clitoral, for vibratory machines continued to coexist with pulsatory machines, designed to automate the process of penetration. Women who did not find the experience of penetration satisfying could enhance their clitoral pleasure by the fantasy of penetration, as embodied in the machinery that automates it. Machinery entered more and more into sexual fantasy, often allowing for the play with forms of forced sex. For part of the force of the pleasure machine is that it allows for the idea of force to be allied with rather than opposed to pleasure. The libinisation of force goes along with the libidinisation of work and what from the mid-nineteenth century came to be known as the 'workforce'. Devices like the sewing machine were drawn into the play of sexual fantasy. Eugène Guibout reported in 1866 that the movement of the pedals in kinetically-operated sewing machines produced considerable genital excitement in certain women, leading in turn to febrility, weight loss, and weakness (Offen 1988: 95). His concern was particularly with the fact that women were required to contribute the motive force to the machine, with vigorous, alternating movements of the legs, which were productive of troublesome friction between them. This seemed the more concerning in that women themselves were both the objects and the subjects of the force. As Monique Peyrière remarks, 'automatic onanism was the danger that assailed the "young woman", who, through the mediation of the sewing machine, could be indoctrinated into unhealthy habits without their volition' (Peyrière 2007: 76; my translation). The problem was not so much mechanisation as such, but rather an incomplete mechanisation that parasitically stimulated and fed upon the energy of the operator. Guibout thought the solution would lie in a machine that made the movement more fully automatic:

> Industry, under the pretext of progress, uses instruments that by themselves engender a demoralization that becomes a cause of ruin for the organism. No doubt, this sewing machine that uses the four limbs of the workingwoman to produce a work in astonishing rapidity, is a wonderful invention. But if this instrument causes the blameworthy effects I have just pointed out to you, it must be given up or modified in the way it works. Another way of making it move, either by steam or by the arms, must be found. (Offen 1988: 97)

In the first volume of his *Studies in the Psychology of Sex* (1910), Havelock Ellis quoted from Thésée Pouillet's *De L'Onanisme chez la femme* (1880) an account of public sewing-machine masturbation in a young woman in a clothes factory, whose feet begin working more energetically, whose nostrils dilate, and who eventually throws her head back with a stifled cry before dabbing her brow and somewhat shyly and shamefacedly resuming her work (Pouillet 1887: 65-6). The most remarkable part of the account is that, as Pouillet was leaving, he 'once again heard, but from another part of the room, another machine start to accelerate its movement. My companion began smiling and remarked that this was so common that one scarcely paid it any attention' (67, my translation; Ellis 1910: 177).

Judith Coffin concludes that women came to be regarded as delicate and unpredictable machines, so that

> [i]t is not surprising, therefore, that images of automated femininity (like women on bicycles or at sewing machines) carried strong sexual associations and that female sexuality seemed so utterly unpredictable and out of control [...] it would have been hard for contemporaries to find a more striking image of the possibilities and perils of modernity than the woman at her sewing machine. (Coffin 1994: 778-9)

Pouillet's description, which is understandably reproduced again and again by cultural historians, seems to have become, as Monique Peyrière observes, a kind of 'primal scene', effecting the entry of the sewing machine into erotic fantasy and literature, alongside the bicycle (Peyrière 2007: 77). At least three fantasy-machines seem to be meshed together here. There is first of all the literal level, at which a woman discovers a mechanism of self-pleasuring in what is supposed to be an instrument of labour. Then there is the ludico-lubricious fantasy offered to the reader of a generation of female wage-slaves single-mindedly working their sweatshops up into masturbation-factories; or, alternatively conceived, a scene of the mass-production of involuntary libido in helpless young women. Finally, if it is final, there is our contemporary fantasy, which includes the panicky, yet pruriently priapic anti-onanist in the mis-en-scène, unknowingly caught up in the works of his anxious excitation. Where the nineteenth-century physician excitedly warns against the delicious perils of feminised machinery, the cultural critic calmly but firmly clicks her tongue, while perhaps also dabbing her forehead, at the absurd but sinister lures of patriarchal professionalism, caught up in fantasy dangers and the dangers of fantasy. At every level, there is a fascination with pleasure (somebody else's) and the pleasure that is a by-product of that fascination. I would not seek to exempt my own account of these interlocked apparatuses from the interlocking even if I knew how. In this imaginary machinery, pleasure is everywhere but in its place.

As has been intimated, the sewing machine was not the only piece of apparatus that required women to pedal vigorously and therefore offered the possibility, or risk, of genital stimulation. By the time Havelock Ellis discussed the auto-erotic uses of bicycles in 1910, the idea was well-established:

> Most medical authorities on cycling are of opinion that when cycling leads to sexual excitement the fault lies more with the woman than the machine. This conclusion does not seem to me to be absolutely correct. I find on inquiry that with the old-fashioned saddle, with an elevated peak rising toward the pubes, a certain degree of sexual excitement, not usually producing the orgasm (but, as one lady expressed it, making one feel quite ready for it), is fairly common among women. (Ellis 1910: 178)

Bicycles have often been incorporated into erotic representations, as may be seen from the examples supplied by the *The Online Bicycle Museum*. ('Bicycle Erotica' n.d.). Georges Bataille's *Histoire de l'oeil* (1928) features a naked nocturnal bicycle ride shared by the narrator and his lover Simone, which culminates in an orgasm which throws Simone off her bicycle, partly perhaps because she has to keep looking round at her partner's erection:

> it was difficult for Simone to see this rigidity, partly because of the darkness, and partly because of the swift rising of my left leg, which kept hiding my stiffness by turning the pedal. Yet I felt I could see her eyes, aglow in the darkness, peer back constantly, no matter how fatigued, at this breaking point of my body, and I realized she was tossing off more and more violently on the seat, which was pincered between her buttocks. Like myself, she had not yet drained the tempest evoked by the shamelessness of her cunt, and at times she let out husky moans; she was literally torn away by joy, and her nude body was hurled upon an

embankment with an awful scraping of steel on the pebbles and a piercing shriek. (Bataille 2001: 30)

Queen's song 'Bicycle Race' (1978) seems to make things pretty clear – 'I want to ride my bicycle / I want to ride my bike / I want to ride my bicycle / I want to ride it where I like' – the song being released as a single backed by 'Fat Bottomed Girls', to which it alludes with the line 'Fat bottomed girls they'll be riding today'. The latter song bounces the allusion back in its refrain – 'Fat bottomed girls you make the rocking world go round' – and the injunction to 'Get on your bikes and ride'. The songs may have been prompted by memories of ample ladies spilling out over saddles in Donald McGill postcards like the one that has the rhyme: 'For a woman these hills are a terrible grind / You see Father's in front, and Mother's behind'.

The genre continues with the Bike Smut Film Festival, which was inaugurated in Paris in September 2011 with the announcement that

> BIKE SMUT is a made-in-Portland anarcho-punk-queer bike collective which, through cyclo-pornographic (or erotico-velocipedal?) films and performances intend to persuade you of the necessity of pedalling and embracing without limits. ('L'érotic Bicycle Film Festival débarque' 2011)

Flann O'Brien's *The Third Policeman* has an extended passage of mock-libidinous celebration of the union of a man with the bicycle he has stolen from a policeman:

> How can I convey the perfection of my comfort on the bicycle, the completeness of my union with her, the sweet responses she gave me at every particle of her frame? I felt that I had known her for many years and that she had known me and that we understood each other utterly. She moved beneath me in agile sympathy in a swift, airy stride, finding smooth ways among the stony tracks, swaying and bending skilfully to match my changing attitudes, even accommodating her left pedal patiently to the awkward working of my wooden leg. (O'Brien 1974: 150)

The owner of the bicycle, Sergeant Pluck, has previously assured the narrator that 'I have always kept it in solitary confinement when I am not riding it to make sure it is not leading a personal life inimical to my own inimitability' (85). This comic sexuality extends into the IT bike, invented by *South Park*'s Mr Garrison, which employs a number of phallic 'flexigrip handles': 'Left side for throttle, right side for steering. The third flexigrip is gently inserted into the anus to keep the driver in place. The final flexigrip is directly in front of the driver so that small switches can be operated with the mouth'.

A prosecution in a Scottish court in 2007 threw up an interesting anomaly. Robert Stewart was allegedly found having sexual relations with his bicycle in a hotel room by cleaners who unlocked his door after knocking several times and getting no reply. Stewart seems to have been prosecuted, not for the offence of sexual congress with a bicycle, but because he did not cease his action when his cleaners walked in on him, instead continuing, with the words 'Well, what is it then?':

> Gail Davidson, prosecuting, told Ayr Sheriff Court: 'They knocked on the door several times and there was no reply.
> 'They used a master key to unlock the door and they then observed the accused wearing only a white t-shirt, naked from the waist down.
> 'The accused was holding the bike and moving his hips back and forth as if to simulate sex.' ('Bike Sex Man Placed on Probation' 2007)

The locution is intriguing here. Simulating sex is one thing; to move your hips *as if* you were simulating sex seems to be the simulation of sex-simulation. At this point, being caught *in flagrante delicto* turns into the distinctively male offence of self-exposure or exhibitionism. This seems reassuringly to restore the sexual order of things, in which men do things and women have things done to them, especially having things done to their eyes by being forced to look at things. (It is not in fact unknown for men to manufacture circumstances in which women, often maids in hotel rooms, are inveigled into coming upon them in the act of masturbating, and indeed this has become an unpleasant genre of internet voyeur pornography.) Rather than bashfully apologising and retreating, the cleaners were shocked enough to report the offence to their employer, leading to the prosecution and conviction. The BBC News account of the case reports that

> Putting Stewart on probation and the sex offenders' register for three years, the officiating Sheriff Colin Miller told Stewart: 'In almost four decades in the law I thought I had come across every perversion known to mankind, but this is a new one on me. I have never heard of a 'cycle-sexualist.' ('Bike Sex Man Placed on Probation' 2007)

The remark seems to belong to the well-established genre of twinkly judicial disingenuousness, for there is a long paronomastic association between cycling and the psychological, with the second term implying, after Freud, the whole panoply of 'psycho-' compounds, this giving rise to the pseudo-malapropism 'trick-cyclist' for 'psychiatrist'. The OED identifies the first usage of the phrase in this sense in Humbert Wolfe's supernatural poetic satire *The Uncelestial City*, which refers to Jim Trodley, a society spiritualist, as 'A trick-cyclist gravely reassembling / The features of the ectoplasmic dead' (Wolfe 1930: 112). If it seems likely that it is the psychic rather than the psychiatric that is in play here, a certain shared association with deceit may be implied. A little later in the poem, a witness to one of Trodley's séances makes the link between the psychic and the psychoanalytic explicit: 'I was one of the *invités chez* Jim Trodley / last night at the ghost-cellar. My dears, too grim, / too psycho-analytically ungodly, / and altogether too divinely Jim' (124). 'Trick-cyclist' suggests the trickiness of the cycling-fetishist, taking their concealed erotic pleasures from the most innocent of actions, while also implying that the psychiatrist is just the same kind of trickster, in the prurient delight they may take in exposing perverse or fetishistic pleasures.

Bicycles are, of course, bathetic kinds of machine, and comedy is never far away when they are invoked. A restrained kind of comedy is similarly apparent in *Sex Machines*, the book of pictures and interviews that journalist Timothy Archibald produced to document his investigation of the makers and users of sex machines in America. The book focuses on devices for providing regular penetrative and thrusting motions rather than vibrators. These are often juxtaposed with household objects, lawn mowers, bicycles, hoovers, or the electric drills used in their construction. The story told repeatedly by the inventors and gadgeteers is of devices adapted for sexual use, often by men with declining capacity for sex, through age or disability. In one case, a husband was disinclined to spend $17 a week to rent a breast pump from the hospital following the birth of his first child so made his own, subsequently adapting it into a sex machine (Archibald 2005: 49). A machine called the 'Huskette' is powered by a $300 food mixer motor, allowing for a poker-faced explication of its dual utility. 'We thought it was silly to spend $300 on something with only one use, so we made this detachable. So, if you wanted to make cookies, you can detach the mixer from the machine, hook it to its original base and bowl – clean it off, of course – and you can get to work in the kitchen' (14). These machines are scarcely imaginary, and everything about the way they are photographed seems designed to reduce fantasy to the minimum. And yet that is precisely the way in which fantasy goes playfully to work in them.

Comedy and the erotic are obviously opposed to each other. The ludicrous nature of many sex machines may be related to the fact that what we often seem to laugh at in sexual matters is the Bergsonian reduction of the free organism to a helpless mechanism. And yet comic pleasure may partly depend on, and even arise from, the mechanical, and the Freudian analysis of the joke work might suggest that the mechanism of comedy can in fact deploy as well as deny libidinous energy. The absurdity of sex-machines may be a kind of alibi, or oblique incentive for what it seems to inhibit. The snigger that is never far away in stories of the sewing machine and the bicycle is isomorphic with the mechanised arousal they promise. Perhaps laughter is like the machine in being something that desire may have to overcome, and in the process intensify itself. For tickling fetishists, being made helpless with laughter is as arousing as pain is for the masochist – indeed, such practices are often known as 'tickle torture' – and perhaps the pleasure has as much to do with the extreme ambivalence of the experience, which resembles orgasm in its condition of passionately self-fulfilling passivity.

Many of the devices described by Hoag Levins in his survey of sexual devices for which patents were applied at the US Patent Office from the mid-nineteenth century onwards indicate this ambivalence, if only because they so often seem concerned with channelling, constraining, impeding, or deterring sexual pleasure rather than enhancing it. Levins describes spiked or electrified anti-masturbation devices, internal penile splints, and a number of ingeniously eye-watering anti-rape devices, including a vaginal harpoon consisting of a spring-driven two-inch spike that would be activated by the pressure of an uninvited penis, and a 'Penis Locking and Lacerating Vaginal Insert', the workings of which are described by its inventor, Alston Levesque as follows: 'the penis may enter this device without great resistance and will activate the blades only upon the attempt of the man to withdraw the penis' (Levins 1996: 207). Yet, if it seems unlikely that such devices could have been turned to the uses of pleasure, one might be instructed by the contemporary popularity of male chastity locks in BDSM play.

There is one very particular mechanical feature of pleasure machines. As Shelley Trower (2012) has shown, the technologies of the late nineteenth century came into communication with a much older discourse concerning the powers of vibration. The poetic language of pleasure has often centred on pulsatory or vibratory forms – tremors, shudders, spasms, quivers, tingles, throbs and, most particularly, thrills, the last-named a metathetic form of the word *thirl*, which means to penetrate or transpierce. In poetic usage, thrills and throbs are strongly ambivalent, conveying suffering and excitement together: Henry More writes in 1642 that he cannot 'to drowsie sensuall souls such words impart, / Which in their sprights may cause sweet agony, / And thrill their bodies through with pleasing dart' (More 1642: sig. A7ᵛ). Thrilling is much in evidence in Gothic fiction, for example in Ann Radcliffe's *The Romance of the Forest* (1791), where we read of the beautiful young heroine Adeline that 'A kind of pleasing dread thrilled her bosom, and filled all her soul' (Radcliffe 1986: 18). Shelley's evocation in *Queen Mab* (1813) of Nero's enjoyment at the destruction of Rome shows the implication of thrilling in the subjection of the subject to their own subjecthood:

> When Nero
> High over flaming Rome, with savage joy
> Lowered like a fiend, drank with enraptured ear
> The shrieks of agonizing death, beheld
> The frightful desolation spread, and felt

A new-created sense within his soul
Thrill to the sight and vibrate to the sound. (Shelley 2004: 187)

To be 'thrilled to bits' suggests the sweet jeopardy of dissolution that is involved in the experience of thrill. The regular, rotary movements of machines have made them a suggestive source of this kind of palpitant excitation. Electrical machines made it possible for such machines to provide sensations at much increased frequencies, which cross the threshold between separable plunges and thrusts and more subliminal impulses, softening, vitalising, or offering to dematerialise the living tissue by turning it into a kind of quivering organ of sound. By simultaneously increasing the speed and reducing the intensity, vibration contains, or hardens, the potentially destructive or diffusive forces of the spasm, at once quickening and restraining. Vibration smooths the discontinuous oscillation of hard quantities (in/out, to/fro, yes/no) into softly continuous quality. It is a minor, subcritical form of the shuddering or shattering force of the spasm, from Greek σπᾶν to draw, tug, or the orgasm, from Greek ὀργᾶν, to swell, which itself seems to be lexically diffused across Greek-derived words in English like *chasm, enthusiasm, plasma, iconoclasm, cataclysm, aneurism,* and *paroxysm*.

The most important duality enacted in the pleasure machine is that between the unchanging, purely repetitive rhythm of the pulsation and the progressive syntax of sexual pleasure leading to consummation. The principles of the hard and the soft are only the material or spatial forms of what in a temporal register is known as duration and variation, or iteration and divergence. I have no need for a record if I can use myself as a recording and playback device, for songs, poems, stories, resolutions, turns of phrase, pronunciations. Repetition is the hardening of time into objects. The pleasure machine provides the unvarying tempo which allows for an ecstatic temporality, for time that transcends temporality. The union of the vibrating mechanism and the climactic organism reproduces the conversion of discontinuous into continuous movement and vice versa that is a central principle of many motors. The vibration repeats itself exactly, like the shuttling of the piston, and this absolute reversibility seems to disallow the movement of time. But it is just this implacable, and wholly dependable, variation without change that seems to permit the convulsive escape from mechanical invariance into duration, giving time an urgent and irreversible direction. The machine arouses the desire to escape from the merely and exactly mechanical. Orgasm is the organism mechanically perfected. It is what allows the organism to come to itself through its self-abandonment, and to do so through a reliably iterable process. The orgasm is never so much a dissolution or forgetting of self that it becomes itself forgettable, and so therefore may be stored as a past with a prospect of future repetition. The mechanoclastic ecstasy is what drives the machine of desire at once onwards beyond mechanism and inwards into its own mechanical workings.

This play between the regular and the spasmodic is enacted within the common slippage – it was one of Mrs Slocombe's characteristic malapropisms in the British TV sitcom *Are You Being Served?* – between orgasm and organism. The two words do not share an etymological root, though they belong to the same techno-scientific register and are connected by the folk-logic that male orgasm can indeed be productive of an organism. The orgasm is thought of in some senses as the proof of the freely self-determining nature of organic existence, exorbitantly in excess of the merely mechanical functions of the body. And yet the orgasm does have an organic function, in that it performs energetic work. Orgasm derives from Greek ὀργᾶν – to swell with sexual desire, be excited or eager, this deriving from ὀργή natural impulse, temperament, disposition, which is comparable to Sanskrit ūrj, ūrjā (nourishment, vigour). Organism derives from Greek ὄργανον – tool, instrument, engine of war, musical instrument, surgical instrument, also bodily organ especially as an instrument of sense or faculty. Its root meaning seems to be 'that with which one works', formed through an ablaut variant of the base of ἔργον, or work. From the mid-thirteenth century, an organ could also mean one who acts as an intermediary, as well as instrumentality in general. The move from *ergon* to *organ* conceals the fact that the organic is in fact originally conceived as the mechanical. They will come together, as we will see a little later, in the term 'orgonics' coined by Wilhelm Reich as a blend of the organic and the orgasmic to describe his vitalistic psychotherapy.

Celibatary Machines

One of the most distinctive of the imaginary machineries of pleasure was introduced by Michel Carrouges in his book *Les Machines célibataires* of 1954. As we have seen, Carrouges's 'celibate machine' was added to the sequence of interlocking machines described in Deleuze and Guattari's *Anti-Oedipus* in 1972. Carrouges evokes the machine as a mythical entity that exists in different variants in the work of a number of early twentieth-century artists and writers. The two principal embodiments of the 'celibatary machine' (as I will call it) are Marcel Duchamp's unfinished *La Mariée mise à nu par ses célibataires, même – The Bride Stripped Bare by Her Bachelors, Even* (1915-1923), often known as *The Large Glass*, and the inscription machine in Franz Kafka's 'In the Penal Colony'. In both cases, Carrouges finds a failure of consummation, but a failure represented erotically: 'Their two great machines have first of all this character of the celibate myth because of the fact that they represent, in a dark and mechanical fashion, the sexual process' (Carrouges 1976: 35; my translation). The purely mechanical is at the opposite end of Lawrentian spiritual *eros*, and yet the mechanism remains haunted and driven by the erotic (37).

In this sense, the celibatary machine can be seen as a myth of demythologisation. Sexual union is the embodiment of the divine will, the sexual and the sacred being embodied in the injunction to go forth and multiply. The refusal of this is a kind of mystical destruction of its mystical force:

> Woman is for man the summit of divine mystery in nature, and the reverse is equally true for the woman. The refusal of love and procreation involves the destruction at once of man, of woman and the presence in them both of mystery. No genuine ecstasy is possible any more. The divine work is fractured, it gives way to a celibatary machine in the process of self-destruction. (47)

For Carrouges, the erotic is both the double and the denial of the sacred. The machine is opposed to the mythological, and yet the mechanical is itself mythical in form and force.

Carrouges introduces his account of the celibatary machine with a comparison of what he regards as its two Ur-examples: the inscription machine in Franz Kafka's 'In the Penal Colony' and Duchamp's *The Large Glass*. There are certain resemblances between the two. Both seem to involve an apparatus divided into two. In Kafka's case, an inscription apparatus communicates the judicial sentence to the convict laid out on the Bed below via the Harrow. In *The Large Glass*, a cylinder-like bride in the upper chamber of the composition appears to communicate, perhaps via some kind of piston system, or perhaps through wireless or radiophonic transmission, with nine 'malic moulds' in the chamber below (Henderson 1998: 193). But they really seem to be associated, not so much by any positive resemblances in the structure or functioning of the machines, as by the fact that the workings of the machine in either case involve a work of writing. Duchamp's machine is in fact a complex assemblage of physical objects, diagrams, instructions, and speculative notes. It is never clear whether these notes, produced over the course of many years, are a kind of explicating 'key' to what is represented in *The Large Glass*, or speculations as to the kind of thing that the work might be or become in various different states of existence. One set of notes, for example, describes, or perhaps prescribes, the effect of an illuminating gas on the nine malic moulds:

> From the top of each malic mould the gas passes along the unit of length in a tube of elemental section, and by the phenomenon of stretching in the unit of length the gas finds itself solidified in the form of elemental rods. Each of these rods, under the pressure of the gas in the malic moulds, leaves its tube and breaks, through fragility, into unequal spangles. Lighter than air (retail fog). (Duchamp 1969: 154)

Duchamp's notes towards his imaginary machine themselves form a kind of formulaic machinery, a sort of imaginary physics in the making. They often seem to prompt or be mirrored in bits of punning writing-machinery, for example in the note Duchamp wrote for the invitation card to Jean Tinguely's *Homage to New York*, an elaborate machine that publicly disassembled itself on 17th March 1960:

> Si la sie scie la scie
> Et si la scie qui scie la scie
> Est la scie que scie la scie
> Il y a Suisscide métallique.
>
> [If the saw saws the saw
> And if the saw that saws the saw
> Is the saw sawn by the saw
> You get a metal Swisscide]. (Duchamp 1973: 170; my translation)

Attempts have been made to reconcile these fragmentary specifications, such as the account of 'The Mechanics of *The Large Glass*' assembled by Arturo Schwartz:

> In the Bride's Domain (the upper half of the Glass), the Bride (1) transmits her commands, in a triple cipher, to the Bachelors (whose expression is the Gas) through the 3 Draft Pistons (6). The 3 Draft Pistons are surrounded by the Milky Way (5), which is the graphic representation of the Bride's three 'blossomings.' Nine Shots (7) were fired at the Glass with a toy cannon expelling a match dipped in wet paint. At the point of impact, the Glass was drilled in correspondence with the nine traces of paint left by the match. This area, known as that of the Nine Shots, is where the Bride's desires meet the expression of the Bachelor's. The latter's desires should have found their graphic expression in a Picture of Cast Shadows (8) formed by the mirrored return of the Sculpture of Drops (9).
>
> In the lower half of the Glass, termed by Duchamp the Bachelor Apparatus, a gas (whose origin is unknown) is cast in the Malic Moulds (11a to 11i) into the shapes of nine Bachelors. The Gas escapes from the Moulds through the Capillary Tubes (12), where it is frozen and cut into spangles and then converted into a semi-solid fog. (Duchamp 1969: 11)

But such synopses are really no clearer than what they attempt to digest. Both Kafka's and Duchamp's seem to be machines whose function is not to work, but to elaborate non-function. They are fictional, or non-frictional, a strange compound of the agitated and the idle, the inert and the impassioned. They are perhaps anticipated by Novalis's 1799 vision of 'a monstrous mill, driven by the stream of chance and floating on it [...] a mill grinding itself' (Novalis 1997: 144), or Conrad's vision, almost a century later, of the 'knitting-machine' of the cosmos (Conrad 1983, 425). When Deleuze and Guattari seize on Carrouges's notion of the celibatory machine they emphasise not so much its mythopoeic character as its refusal of the principle of production, in favour of intensities of feeling:

> The question becomes: what does the celibate machine produce? what is produced by means of it? The answer would seem to be: intensive quantities. There is a schizophrenic experience of intensive quantities in their pure state, to a point that is almost unbearable – a celibate misery and glory experienced to the fullest, like a cry suspended between life and death, an intense feeling of transition, states of pure, naked intensity stripped of all shape and form. (Deleuze and Guattari 2000: 18)

The strange career of the celibatary machine does not end with Deleuze and Guattari. Michel de Certeau's *The Practice of Everyday Life* adopts the phrase to describe the passivity of the 'consumer-sphinx', on whose body the law of authoritative discourse inscribes itself:

> The television viewer cannot write anything on the screen of his set. He has been dislodged from the product; he plays no role in its apparition. He loses his author's rights and becomes, or so it seems, a pure receiver, the mirror of a multiform and narcissistic actor. Pushed to the limit, he would be the image of appliances that no longer need him in order to produce themselves, the reproduction of a "celibate machine". (de Certeau 1984: 31)

Carrouges's machines, found in the work of Jarry, Roussel, Duchamp, and Kafka, are the 'myths of an incarceration within the operations of a writing that constantly makes a machine of itself and never encounters anything but itself' (150). De Certeau assures us that '[c]elibacy is scriptural' (151). And yet some real work does seem to be done by this self-signifying machinery, this empty eroticism. Is a machinery of pure writing simulating a machine, or dissimulating the fact that it is not really any kind of machine at all, a machine or not? De Certeau nevertheless finds in it a kind of force, a rapture in its power of suspension:

> It is through this stripping naked of the modern myth of writing that the celibate machine becomes, in a derisive mode, blasphemy. It attacks the Occidental ambition to articulate the reality of things on a text and to reform it. It takes away the appearance of being (i.e., of content, of meaning) that was the sacred secret of the Bible, transformed by four centuries of bourgeois writing into the power of the letter and the numeral. (153)

If this is a machine of pleasure, it is a pleasure that consists in the absolute ambivalence as to whether there is any celebration to be had from the celibate machinery.

True to their unfixed, non-consummatory nature, celibatary machines have begun to multiply in different times and places. Roy Caldwell finds a bachelor machine at work in, indeed, identical with, the 'text-machine' that is Laurence Sterne's *Tristram Shandy*, at once sterile and yet gloriously fertile in forms of impotence. Here 'the text-machine turns parricide, bastardizing itself, establishing the conditions of its bachelor solitude' (Caldwell 1993: 111). Branka Arsić tracks out the workings of the celibatary machine in the failure, passivity, and uncertainty of figures like Bartleby the scrivener, who, in refusing the machinery of writing, creates a sort of machinery of non-writing, or unwriting:

> Bartleby must become the closeness and fullness of a cipher, which is to say not only that Bartleby stops writing, but also that he stops being written, that he escapes the inscription of the law upon his body and thus the possibility of being read by it. He is neither written nor readable. He is the sole spectator of what he sees but what he sees is his own solitude; he is thus the endless circulation of a text within itself with no escape from itself. (Arsić 2005: 98)

In these kinds of evocations, actual machines, with anything at all in them of material forms or actions, energies or resistances, seem to have melted away entirely. And yet that may be the very source of the pleasure attached to this playful machine-that-is-not-one. The pleasure of the celibatary machine lies in its infinite complication and deferral of pleasure. This puts it in obvious opposition to the other kinds of pleasure machine, like the vibrator, which seem by contrast both literal and wedded to the actuality of pleasure as defined and required outcome. In the one case the machine creates pleasure through indefiniteness; in the other through an automation which makes it reliable, absolute, and, as with the fantasy of the frantically pedalling seamstress, almost autonomous of the subject, who 'pleasures' himself or herself, or is 'pleasured' thereby. To 'pleasure' somebody sexually is a usage that, while not previously unknown, becomes much commoner in the second half of the twentieth century, in which devices of various kinds have made the commerce of subject and object much livelier, if that is entirely the word.

Orgasmechanics

In Woody Allen's *Sleeper* (1973), the hero finds himself in the world of 2173 in which person-to-person sex has been replaced by a form of machine-mediated congress that takes place inside a booth-like apparatus called the 'orgasmatron'. Allen has in his sights an apparatus known as the orgone box, or, more precisely (well, more 'precisely'), the 'orgone accumulator' that was invented by Wilhelm Reich soon after his arrival in the US in 1940. Reich claimed that the box, which consisted of several alternating layers of 'organic' material (wood) and inorganic material (metal), insulated with rock wool, was a kind of battery or accumulator for orgone, a cosmic life energy which was spread throughout the universe. Orgone seems like a close relative of the ether, which was thought by some similarly to be both spread everywhere in the universe, and yet capable of being condensed or concentrated (Connor 2010: 80). In the case of the orgone accumulator, orgone was attracted and absorbed by the wood, but radiated inward by the metal, allowing it to be concentrated on the inside of the box, causing, among other effects, an increase of temperature (why, once it had got into the box, it could not get out again, nobody seems to know). It occurred to Einstein, whom Reich tried to convince of the authenticity and virtues of his apparatus, that, if this were all that were needed to create heat energy, and there were only accumulation without dispersal, Reich would in fact have devised a 'great bomb' (Reich 1999: 199). (Associates of and commentators on Reich have glossed 'great bomb' as 'an enormous impact on physics', but there is no reason not to take Einstein's comment literally). Sitting in the orgone box was supposed to restore depleted levels of this vital stuff, leading to fuller and more consummate orgasms, which Reich believed themselves had powerfully vitalising effects, as well as helping to cure various kinds of illness, most especially cancer.

The orgone box was adapted enthusiastically by writers like Norman Mailer, J.D. Salinger, Paul Goodman, Allen Ginsberg, Jack Kerouac, and William Burroughs, and became a centrepiece of the politics of sexual liberation of the 1960s. One of the attractive features of the orgone accumulator was that it could so easily be made at home, from cheap and familiar materials (William Burroughs knocked up one for himself). Instruction manuals for their construction are still available: one advising that accumulators should be built in rectangular or cylindrical shapes, since 'accumulators made in the shapes of cones, pyramids, or tetrahedrons have yielded occasional inexplicable life-negative effects' (DeMeo 1989: 44).

Following investigation by the Food and Drug Administration, Reich's machinery was condemned as fake, and in March 1954 he was ordered to destroy all his accumulators. When one of his associates sent an accumulator through the mail, Reich was sentenced in May 1956 to two years' imprisonment for contempt of court, and died in prison of a heart attack. This ill-judged over-reaction by the authorities confirmed Reich's status as a hero of the counterculture, and the purveyor of a great liberating truth suppressed by the dark powers of the Establishment, and his ideas about orgone have continued to hum away in popular culture and music. There is a funk-revival band called Orgone, and another kind of orgone-concentrating Reichian machine was at the centre of Kate Bush's song and video *Cloudbusting*, which begins with the line 'I still dream of Orgonon', the name of Reich's house in Maine, suggesting some equivalence to Mandalay. An orgone accumulator van, which perhaps can be hired for mobile orgonifying sessions, is regularly seen parked a couple of streets away from where I live in London.

The orgone box emerges out of a long preoccupation with the relations between sexuality and machinery in Reich's writing. Much of this is documented in the three editions Reich published of *The Mass Psychology of Fascism*, the first two in German in 1933 and 1934, the third in an English translation from Reich's revised manuscript in 1946. At the heart of Reich's theory of fascism is the view that it is formed from a conjuncture of mysticism and mechanism, a conjuncture that in fact anticipates his own theories and machineries of sexual energy:

> '[F]ascism' is the basic emotional attitude of man in authoritarian society, with its machine civilization and its mechanistic-mystical view of life. It is the mechanistic-mystical character of man in our times which creates fascist parties, and not vice versa. (Reich 1946: ix)

There are two sides to this mechanism-mysticism pairing. First of all, there is a mechanism of substitution effected by what Reich calls the 'biopsychic apparatus' (117). Reich has recourse to a fairly conventional Freudian hydraulics in evoking the ways in which baulked or disavowed 'orgastic longings' are redirected into other channels. The combination of desire and prohibition makes these feelings seem both intense and occluded:

> The continued tension in the psychophysical organism is the basis of daydreaming in the child and adolescent which readily continues in the form of mystical, sentimental and religious feelings. This characterizes the atmosphere of the mystical, authoritarian individual. In this way, the average child acquires a structure which cannot help but absorb the influence of nationalism, mysticism and superstition of any kind as avidly as a dry sponge absorbs water. The reaction of the biopsychic apparatus is the same when it reacts to gruesome fairy tales, later to mystery thrillers, to the mysterious atmosphere of the church and, finally, to militaristic and nationalistic display. (116-17)

This displacement is itself a kind of mechanism, in that it involves unconscious processes: 'sexual suppression serves the purpose of mechanizing the masses of people and making them dependent' (184). Reich notes the contradiction between technology and mysticism, which is being combated in Russia: 'for the first time it becomes possible on a mass scale to replace religion by natural science, to replace the illusory protection of superstition by technical achievement, to destroy mysticism by sociological elucidation of the function of mysticism' (148).

But scientific rationality is not enough. Reich calls for a kind of rational mysticism in the form of 'full sexual consciousness'. This would involve more than intellectual understanding: 'Such ideological manipulation is explicitly avoided, since it would not alter the biopathy. The process is, rather, that of unmasking the mystical attitude as an antisexual force and of liberating the energies which nourish it, of making them available for rational use' (156).

So there was a mechanism of repression, which produces mystification – 'If I talk with a sexually inhibited woman in my office about her sexual needs, she will mobilize her whole moral apparatus against me, and I shall not be able to convince her of anything' (160). But the therapeutic response to this mechanised mystification, or mystified mechanism, is also a kind of mechanical intervention, rather than an act of interpretation, Reich having always been frustrated and unconvinced by the orthodox Freudian emphasis on interpretation. Reich's term for his mode of analysis, 'sex-economic therapy', implies a kind of mechanical adjustment of energies in some more directly corporeal wise than Freudian analysis.

Reich's valuing of the orgasm was itself intensely organicist. Like many another modernist – and like many fascists – he opposed the mechanism of modern life, writing at length in of the ways in which humans have become machine-like, even to the point of understanding their own biological functioning in mechanical terms:

> *In the course of thousands of years of mechanical development, the mechanistic concept, from generation to generation, has anchored itself deeply in man's biological system. In so doing, it actually has altered human functioning in the sense of the machine-like* [...] Man not only believes that he functions like a machine, *he does in fact function like a machine.* He lives, loves, hates and thinks like a machine. (293)

Reich insisted that the 'functional' laws of the orgone were completely different. 'The specific biological energy, the orgone, follows laws which are neither those of mechanics nor those of electricity. Because man has been biased in favor of a mechanistic concept of the world he has been incapable of grasping the specifically living, nonmechanistic functioning'. (287-8). He urges that 'sex-economic mass psychology' (184) must emerge 'organically': 'one cannot "think out" or "plan" a new order; it must grow organically, in closest contact with the practical and theoretical facts of human life' (181). Political identification cannot be mechanistic, for those who 'identified themselves, economically and ideologically, with this or that party machine' were 'rigid and inaccessible to any new insight' (182).

But the organic hypothesis is just as mechanical, and as mystically so, as what it aims to surpass. 'The machine' is on both sides of the subject/object dichotomy for Reich. It is the object of his analysis, but also its governing principle. Reich sees mechanism as itself a kind of machine, and his own explication of the machinery of mechanist thinking is powered by the very same machinery. Condemning the Prussian 'mystical machine-man' (295), Reich becomes hopelessly entangled in his own diagnosis here, his own metaphysics of life appearing to be just such a mixture of the mystical and the metaphysical as he analyses. He seems to come close to recognising this when he writes of the way in which the machine mysticism of fascism is itself driven by the longing for the freedom of the organic: 'These diverse cries for freedom are as old as the machine-like aberration of the human plasm' (296). Indeed, the very fact that his analysis of the relation between the mechanical and the metaphysical in fascism is in many ways so acute may be what renders him so susceptible to the same paradoxical logic of mystico-mechanical anti-mechanism.

The orgone box emerges out of a thirty-year campaign on Reich's part for a politically emancipatory form of psychoanalysis centred on the lifting of sexual repression. As already noted, Reich seems to have been propelled into his theory of the orgasm by an impatience with the indirectness of psychoanalytic method and theory, and a wish to intervene more directly and corporeally in the illnesses of his patients. In this, he moves in the opposite direction from Freud, for whom the theory and method of psychoanalysis came about as a consequence of abandoning direct physical interventions like hypnosis – Reich was very impressed by Freud's speculations during the 1890s that the libido might be electrical or chemical in nature (Turner 2011: 173). But, having moved from the mental towards the physical – and, as Rachel Maines has shown, he was not the only physician offering therapeutic orgasm to his patients in the early twentieth century – Reich then substituted a kind of metaphysical physics for that literal intervention. Reflecting on what he had learned from Freudian psychoanalysis in *The Function of the Orgasm*, Reich wrote

> It was entirely without conscious premeditation that I used the simile of electricity and its energy. I had no idea that sixteen years later I would have the good fortune of demonstrating the identity between bioelectric and sexual energy. Freud's consistent, natural-scientific thinking in terms of energy captivated me. It was objective and lucid. (Reich 1974: 30)

Looking back on his treatment of schizophrenic patients, Reich saw paranoid delusions of having one's thoughts or body controlled by external agencies as a representation of fact:

> the patient might have the feeling that he is being electrified by a secret enemy, whereas he is merely perceiving his own bioelectric

> currents [...] These observations formed the basis of my later conviction that the schizophrenic's loss of a sense of reality sets in with the misinterpretation of his own burgeoning organ sensations. All of us are merely a specially organized electric machine which is correlated with the energy of the cosmos. (42)

So, with the suspension of customary ego boundaries, the mental patient 'at least has an inkling of what the cosmos is' (43). Reich found in the work of Victor Tausk, whose essay on the influencing machine he encountered soon after it was published in 1920, an anticipation of his own infatuation with mystical machinery:

> [Tausk] demonstrated that the apparatus which influences the schizophrenic is a projection of his own body, in particular of the sexual organs. It was not until I discovered bioelectric excitations in the vegetative currents that I correctly understood this matter. Tausk had been right: It is his own body that the schizophrenic patient experiences as the persecutor. I can add to this that he cannot cope with the vegetative currents which break through. He has to experience them as something alien, as belonging to the outer world and having an evil intent. The schizophrenic merely represents in grotesquely magnified condition what characterizes modern man in general. Modern man is estranged from his own nature, the biological core of his being, and he experiences it as something alien and hostile. He has to hate everyone who tries to restore his contact with it. (46)

Late in his life, Reich recapitulated the partial return that Freud himself had made to biophysical explanation in *Beyond the Pleasure Principle*. Reich had doubted the existence of the death drive for more than 30 years, perhaps because, for him, the identification of the organic with the orgasmic meant that there could really be no alternative to the pleasure principle except the denial or repression of pleasure. But, in the last essay he ever published, Reich was willing to identify the death drive with the Deadly Orgonic Energy (DOR) he believed he had discovered in the atmosphere above desert regions, which thwarted living growth, and encouraged the growth of unhealthy botanical forms like cactus, the spikiness of which resembles the aggressive character armour of the orgastically-congested neurotic. This energy could be dispersed by 'cloudbusting' apparatus consisting of arrays of hollow tubes which channelled positive atmospheric orgone against the necrotic clouds. Reich saw DOR as responsible for the growth of cancer, and even proposed that Freud's formulation of the theory of the death drive in 1920 was an intellectual intimation, like that of the schizophrenic sensing the electricity in his own body, of the negative energy that was already at work in his system forming the cancer of the jaw that would be diagnosed in 1923 (Reich 1956: 7; Gramantieri 2016: 217). Reich's inability to tolerate the difference between things and their mental representations, characterised by Freud in 'The Unconscious' (1915) as 'the predominance of what has to do with words over what has to do with things' among schizophrenics (Freud 1953-74: 14.199), meant that even the idea of negative energy was to be regarded as exercising its force ('speak of the devil'). Reich warned himself and others repeatedly against the mysticism that he believed was the effect of blocked organic energy. But his defence against the flight into the occult was a refusal of metaphor or abstraction and a physical literalism that was itself a perverse kind of mystification. It is perverse precisely in its insistence on an absolute and ideal regularity of effect and an absolute congruence amounting to identity between idea and actuality. The mysticism of fascist mechanism took the form of an occulting of proof or evidence; Reich's hyper-rationalised mysticism took the form of a denial of any approximation or play in the system connecting conception and actuality. Actual machines always have to struggle against resistance and impediment: occult or ideal machines can be guaranteed to function entirely logically and to overcome every resistance. For the believer in mystical truth, it is essential that things do not add up: for the superstitious adherent of a rational system, there can be nothing that does not add up – the occultism of the *ecco là*.

The efforts to defend against fantasy become ever more fantasy-driven. In one sense the mechanism of Reich's system is an attempt to make it immune from any criticism – none of this is any of my doing, Reich seems to say, it is just part of the cosmic machinery, of the way things fundamentally are. This makes Reichian orgonomics a reliable source of imaginary pleasure, the pleasure of seeing reality entirely subjugated to imagination. In another sense, Reich, who must have known perfectly well how closely his delusions resembled the systematic delusions of psychotics, may in fact have felt the pressure to maintain or even manufacture difficulty or impediment, in the form of that pseudo-complexity that is never far away in the psychography of all machines. There may even have been the temptation to push his system further even than seemed necessary into absurd hyper-consistency, precisely in order that he, and his machinery, for all their ludicrous and wish-fulfilling simplicity, or perhaps precisely because of them, could continue be misunderstood.

Ironically, Reichian organicism and the aesthetic-libidinal rage against the machine of conformity it seemed to underwrite seem to have been taken up into what, a few years after Reich's death, Herbert Marcuse would call 'repressive desublimation', in which all the aesthetic-libidinal energies allegedly released from repression in fact 'become cogs in a culture-machine which remakes their content. Artistic alienation succumbs, together with other modes of negation, to the process of technological rationality' (Marcuse 2002: 68). In *Eros and Civilization* (1955), Marcuse had criticised the 'sweeping primitivism' and 'wild and fantastic hobbies' of Reich's later years. (Marcuse 2005: 239). Since the pleasure principle has been made identical with the reality principle, as much in sexuality as in art, '[i]t appears that such repressive desublimation [...] operates as the by-product of the social controls of technological reality, which extend liberty while intensifying domination' (Marcuse 2002: 75-6). From Marcuse's perspective, machineries of orgastic enlargement and emancipation are expressions of a 'rational universe which, by the mere weight and capabilities of its apparatus, blocks all escape' (75). For Marcuse, the 'mobilization and administration of libido' (78) blocked the very possibility of the transcendence of the socially given it seemed to promise.

It has seemed to human beings for most of history that the function of machines ought to be to save us from performing labour or allow us to perform labour that would otherwise be beyond us, thereby leaving us free to enjoy a life of leisure. From the late nineteenth century onwards, the development of machines that mediated not only the kinetic powers of the human body, but also sensory experience, has brought machines into ever closer proximity with pleasure. In a world in which pleasure is everywhere mediated through machines rather than arising spontaneously from the labour-saving function of machines, pleasure has become more and more intricate and laborious. We now have to work as hard at the taking of pleasure, and at managing the byproducts of its abundance, like addiction, as we used to at overcoming the impediments to desire. But this may not be entirely unprecedented in human history, and perhaps machinery has simply moved into the mediating place that the imagination of machines always occupied.

5
Medical Machines

Human beings use equipment to engage with and transform the world. But, almost as soon as humans began making knives and spades, they also began making needles, combs, brushes, strigils, and other instruments for the examination, care and transformation of the body. Such instruments are often very intimately twinned with the body, to the point of becoming what are well-named as 'personal effects', which are so closely identified with a particular person that they may be interred with them after death. Medical machines are usually more elaborate than instruments, and may be operated by others, who become part of the procedure.

There are no machines that come closer to the body than medical machines, and there are no machines which exhibit a closer analogy with the imagined body itself. The logic of many medical machines seems to be that the best way of working on the body to repair injury or cure disease is not to correct or supplement it but rather to act with it according to its own principles. Medical machines, real or imagined, often function as visible allegories of the working of the body itself – machinic embodiments of a theory of the body.

To a very large degree, medical machines need description. This is for several reasons. One is that medical machines often involve the patient having to subject themselves to some kind of mechanical operation. This is not always or necessarily the case with other kinds of machine. If I am shown a pulley that works better than the one I have been using for the last twenty years, I will very likely adopt it without requiring an exposition of the mechanical principles it on which it depends. I may be professionally interested in knowing how and why a machine gun fires many more bullets than a rifle, but if my aim is to kill large numbers of enemy soldiers, this knowledge is not really necessary to win my credence in the efficacy of the machine or consent to use it.

But a medical machine is almost certainly going to be used on me in some way, and I may very well be asked to submit to some discomfort as part of the process. I am not going to work a medical machine, it is going to work on me, and I am going to be worked, maybe even worked over, by it. Rarely is this just an inconvenience. For I will often need willingly to work with it, to subject myself to its operations.

'I sing the body electric', Walt Whitman famously sang in 1855 (Whitman 2004: 127). What then follows in Whitman's poem does not seem to have anything very much to do with electricity, except in occasional metaphorical hints thrown out for the reader: 'charge them full with the charge of the soul [...] Mad filaments, ungovernable shoots play out of it' (127, 130). Instead the poem is a celebration of 'the exquisite realization of health' (135) which finally affirms that 'these are not the parts and poems of the body only, but of the soul, / O I say now these are the soul!' (136) It is altogether a rather creepy performance. There might in 1855 have been many more literal ways in which Whitman might have not just sung the praise of the body adjectively electric – electric because suffused with and expressive of soul – but have electrified the body, by singing it into an electrical condition. The story has by now frequently been told of how human bodies were more and more routinely subject to the operations of electricity during the nineteenth and early-twentieth centuries. But this process was a matter of singing, or at any rate, of writing, as much as of electrical actions. Medical electricity comes into being as a 'cultural technique' or *Kulturtechnik* (Siegert 2015), a technology that is not just brought to bear on the body, but itself embodies a technesis: a set of techniques and a corporeal rhetoric which encompasses lyric poetry, technical reflections and, perhaps most powerfully of all, and sometimes conjoining the two others, advertising copy.

Medical procedures are known as 'operations' and these operations occur in places called 'theatres'. That is, they are operations that are to be performed in the sense of being represented, or acted out as well as being simply enacted. There is something implicitly demonstrative and pedagogic about every operation, however simple, whether the taking of a pulse or the giving of an injection. It is an important part of the procedure that patient and doctor are both aware of and compliant with the fact that it is a procedure – and this is true of most cultures in which medical operations take place. This is not just a matter of ritual, since there are many ways in which the safety or effectiveness of medical practice depends on carers being trained to follow protocols rather than to exercise their individual judgement. The word *protocol* derives from Ancient Greek πρωτο- first + κόλλα glue and refers to the first or authenticating leaf of a papyrus. Nowadays it no longer means original authority, but rather, as in *internet protocol*, an agreed practice, that is confirmed in its repeated performance. A medical protocol is a kind of mechanism, a way of preventing variation in performance.

The history of the word *operation* shows how the work or the procedure also gradually becomes further and further mediated by mechanism. Its earliest citation in this sense in the OED is a 1425 translation of Guy de Chauliac's *Grande Chirurgie*, in which the 'operaciouns of cirurgiens' are described as 'hande crafts' (Chauliac 1971: 4, 2). This is really a tautology, for chirurgy, the origin of the word surgery, is from Greek χειρουργός, which, conjoining χείρ, or hand, and ἔργον, or work, means, simply, handiwork. It is hard to believe that any kind of incision could be made without the aid of an instrument, and indeed Chauliac says that '[t]he instrumentis or cirurgiens with pe whiche pese pinges beep fulfilled beep manyfoold' (4). The word operation quickly starts to mean some formalised procedure, or 'artificialle and normaticke applicatione, which is brought to passe, and wrought by the handes, one mans bodye' as it is described in Jacques Guillemeau's *The Frenche chirurgerye; or, All the manualle operations of chirurgerye* of 1598 (Guillemau 1598, fol.1ᵛ). Gradually, the various surgical procedures, often identified with the names of their originators, were generalised, both by being externalised and made available for others through the development of technique (which in English often signals the bodily internalisation of what in German is called a *Technik*), and through the mediation of various instruments, which themselves have to be mediated by techniques of use. For a century and a half after the development of general anaesthetic the patient was usually unconscious during an operation. Before that period, to be unconscious during a surgical procedure normally meant that you were in fact dead. There are very great advantages to the development in recent decades of local anaesthesia, combined with much more localised microtechniques, even and especially for very serious interventions like brain surgery, for patients may be able to provide valuable feedback during such procedures that they could not when unconscious. The spread of such practices involves a further extension of the involvement of the patient in the development of shared social techniques and comportments. Despite routine complaints about the objectifying effects of medical technologies and procedures, patients are rarely passive, since they have a very considerable, and growing, part to play alongside medical professionals in 'performing operations'.

If there is one general principle that is thought to distinguish mainstream from alternative modes of therapy, it is the idea that the first kind of medicine is alienating, focussing simply on the illness or affected part, where the second is 'holistic', in addressing itself to the whole patient. '[M]odern medicine must come to terms with the whole person as a social, cultural, and spiritual being, rather than treating the body as a senseless automaton and the mind as a meaningless illusion', Richard Blaustein enjoins (Blaustein 1992: 39). But the dichotomy between automaton and whole person is a false one. Nearly all medical practices, traditional and modern, alternative and mainstream, are simultaneously alienating and holistic, and indeed in both cases the two perspectives are necessarily closely co-ordinated with each other. Whatever being 'holistic' means, it will always in fact involve a subtraction. For one thing, anybody who conceives themselves as being ill (by no means a spontaneous or straightforward matter) will by that very token be viewing themselves through a kind of synecdoche, or taking the part for the whole, in that their illness will mean that some particular portion of their physical being has assumed a kind of autonomy that it does not have when they are well. I cannot have a toothache without sequestering my tooth from my general experience in a way that is out of the ordinary. To be ill is to be uncomfortably constrained to take some part for the whole – the more I am, rather than merely having, a fever or headache, the iller I am. This is particularly the case when I am constrained to pay attention to the functioning of some part of me – my heart or kidneys for example – that ordinarily can be relied upon to perform their functions unsupervised, and indeed depend upon it. To be well is not to have to pay this kind of selective attention. We are sometimes enjoined by proponents of meditation and relaxation techniques to pay close attention to our breathing, but insofar as breathing is designed to be an autonomic function, and in fact must be if we are to survive, to pay attention to your breathing is, in the terms I am proposing here, literally sick.

The process of being ill usually involves much more than having a fever or headache: I may be uncomfortable, but I am not really ill until I have formed the view that I need to take counsel about my condition, even if it is with myself, and this of necessity requires that I focus on some part of my physical existence (for the simple reason that you cannot focus on the whole of your physical being). And, if being ill involves, and perhaps to a very large degree actually is, the exercise of selective attention, then the kind of medical attention that may be invited or requested is very likely to accord with this perspective, at least to begin with. Of course, it is possible, and indeed increasingly likely, that the kind of expert attention I seek will identify some larger or less localised cause for my headache (dehydration, high blood pressure, unhealthy radiation from overhead cables, spirit possession, tumour). It is often the role of the medical professional to assist in this apparent enlargement of explanatory context.

In fact, however, no matter how apparently enlarging it may seem to be (your simple headache is in fact a sign of industrial pollution/cancer/trauma/bewitchment), every 'holism' will in fact involve a necessarily drastic abridgement of the potentially infinite number of ways I might have of conceiving myself and my relation to the world. This is because in almost all cases what counts as medical explanations will move from the relatively unordered to the ordered. That is, they will substitute for my ordinary way of conceiving my physical being – which is in fact not to have much conception of it at all – a more systematic understanding, which feels as though it is more comprehensive than what I started with, but will do so precisely because, and always to the degree that it is, a kind of reduction. All such systems derive what authority they have from being mechanistic, which is to say coherent. No homeopathic textbook has ever said that the dilution of an active ingredient sometimes increases its potency, and sometimes makes it toxic, in a kind of random variation, though that would be a genuinely non-reductive way of thinking of things. Homeopathy is reductive insofar as it is an explanation applying consistently to a set of circumstances that themselves operate consistently. There is no real chance of choosing between magic and mechanism, since the magical is almost entirely mechanistic (this is one of the reasons that James Frazer preferred magical thinking to the religious thinking that he believed succeeded it in human history, since magical thinking is at least internally coherent). One may of course have a choice of mechanisms, and it may be better to choose more complex rather than simpler explanations: but it does not look, thank goodness, as though human beings are going to develop an appetite for genuinely non-mechanistic explanations any time soon. We should recognise as well that the work done by all the things that count as medical explanation depends upon an interplay of simplicity and complexity, in which a simple explanation that integrates a great deal of apparent noise or complexity will always do more work than an explanation that seems to earn its simplicity through the setting aside of complexity.

The extension of different kinds of machinery and mechanical operations from the seventeenth century onwards has meant that these different kinds of conceptual mechanism began more and more to interact with the actual forms of equipment that came to play an enhanced role in medical encounters. We have seen in the relation between Reich and Freud a particularly dramatic tension between conceptual and physical forms of machinery, Freud insisting that the libido was not an isolable stuff or quantity, Reich declaring that it not only existed in the form of orgone, but could be observed in the form of the dancing particles he called 'bions' (Reich et al 1938). But perhaps this interplay is always at work in medical encounters, and is very often mediated through different kinds of discursive machinery – the techniques-technologies of writing, speaking, speculation, and social action that constitute technesis.

So medical machines must always be, if not in fact imaginary, then always subject to a work of imagining, which will both be effected by and will itself maintain in being an assemblage consisting of the apparatus, the way in which it is deployed by its operator, the theory of its operation, the body of the patient, and the patient's response to the elements of this ensemble (to name only these). Any particular medical instrument or machine will always need to operate within this co-ordinated machinery of mediated imagining. This is particularly the case where investigating subjects experiment on, or with themselves, as was often the case with galvanic experiments, for example in the work of Alexander von Humboldt and Wilhelm Ritter: in such cases, as Joan Steigerwald notes, '[b]y incorporating their own sensory selves into the experiments, not only was both the subject and object affected by the tools of inquiry, but these instrumentalized objects and subjects in turn affected the tools and methods of investigation' (Steigerwald 2016: 110).

Every machine has as part of its use the demonstration of its use, all machines being in that sense kinds of model. In the mechanically-mediated medical encounter, models and machines exchange qualities with particular intensity: the machine is a model for how the body is taken to work, that model always involving the production of the body as some kind or other of machine. The workings of that machine will always include its potential to be modified or affected in various ways by its environment, which must necessarily include the therapeutic environment constituted by the machinery of the encounter. The theory of the body's machinery must include an image of how the body will function within therapeutic operations, a theory that will script, not only the body, but the corporeal-conceptual rhetoric of the therapeutic operations themselves. One can see in the operation of many forms of medical machinery the construction of an automaton, or simulacrum of the body. Indeed, the interest in automata provided an important form of interface between the rather lowly arts of medicine and mechanics in the early modern period (Bondio 2009).

So the principal function of many medical machines may be symbolic rather than actual. Or, rather, we might say that their function is to provide a way of making the symbolic actual and practical, a way of making a particular embodied image or idea of how the body is taken to work itself perform work on the ailing body. It is a medium and a machine at once, that both signalises the condition of the body and provides a way of intervening in its condition to alter it. One of the reasons that electricity became during the late eighteenth and nineteenth centuries not only the power-source, but also the master-image of the body as a network of internal and external flows, is because electricity had been from the very beginning a means of exhibition and proved so quickly and powerfully adaptable through the nineteenth century to the work of communication and signalling.

It is often said that the encounter between doctor and patient has become ever more abstract and mediated because of the incursion into that relationship of machines. In the case of computer-simulated diagnostic exchange, indeed, an automated procedure may take the place of a human doctor altogether. Many forms of alternative therapy may seem, or at least claim, to offer by contrast a more immediate and therefore more 'natural' or 'human' encounter, in which the one providing the treatment will rely upon a kind of direct bodily communication, often involving the forensic intuition of the hand rather than some supplementing instrument. In fact, however, there is always the interposition of some mediating mechanism or other. The hand, for example, may not in fact be laid directly on the body of the patient, but rather, as in some osteopathic or chiropractic performances, be held at a distance from the body, precisely to allow for the mediation of some imaginary force or vibration, emanating from the body. At the beginning of his career, Freud used to act out a kind of mesmeric relation with his patients by pressing his hand against their foreheads. When he abandoned this 'small technical device' (Freud and Breuer 1991: 270), thereafter relying on the operations of the talking cure, he performed a different set of relations between the patient, the physician and the patient's body, in which the more abstract techniques and interpretative apparatus of psychoanalytic theory itself formed the necessary connective medium. This kind of medium is always at work, in the protocols that govern every kind of medical contact, and these protocols will almost always involve dynamic systems of desire, risk, danger, anxiety, excitement, and care. Whatever curative method or force may be employed, it is often important for it to be applied in carefully regulated doses, for example, in order not to overload the patient's system. The Goldilocks routine of precise adjustment between pathology and treatment is part of the performed relation of careful concern that guarantees that what is happening is medical.

If there is always mediation by some kind of machinery, whether material or conceptual, however immediate it may appear to be, then we must also say of the mechanically-mediated encounter – involving probes, monitors, drips, clamps, catheters, investigative instruments, and so on – that it in fact always constitutes a more rather than a less complex entanglement of different kinds of system, and is therefore a cohesive rather than alienating mediation. There can be no relation without mediation, and everything, whether in conventional or alternative medicine, always happens in some kind of middle, to borrow Bruno Latour's formulation (Latour 1993: 37), a medical middle that is constituted whenever there is any kind of encounter that will count as curative, therapeutic, or consultative. The doctors of the insane who began to be called *alienistes* in mid-nineteenth-century France were called so not because they alienated the insane, through their treatments or their regimes of incarceration, but because they treated those who were *aliené*, put at a distance by their insanity.

Mechanisms are mediations, and for that reason require mediating substances, objects that can be regarded in the light of powers, potentials, or forces. These substances are surprisingly long-lived and capable of being adapted to new circumstances and of meeting new needs. The humoral theory that governed medical thinking, not just in Europe but in many other parts of the world, gives special status to four fluids produced in the human body: blood, phlegm, choler, and black bile, some of which may seem more literal than others, but all of which were and are in fact imaginary – that is, what they are is a function of what they are imagined to be capable of doing in relation to the others. Blood exists, but humoral blood has to be existed as part of a system of ideas. The etymological origin of the Latin *umor*, moisture, is not very certain, though it seems to have involved an analogy between moistness in living beings and the sap or juice of plants. An association has also been suggested with Lithuanian *umus*, quick-tempered, or hasty, while the initial 'h' probably derives from the false idea that the word is related to *humus*, soil (also associated with the word human).

The humour had a dual function. On the one hand, there were the physical functions performed by the different human fluids. But the humours also had the symbolic function of carrying the understanding of the human body as consisting of static and dynamic components – of substance and passage. A body is something closed off from its environment, that which uniquely occupies one area of space and can occupy only one such area of space at any one time. But a body is also that through and across which things flow: cut it and it bleeds, kill it, and it in effect dissolves. Bodies are engines, and in this perhaps are like every other kind of engine, for the systematic regulation of flows and constraints. Bodily fluids are the means of communicating, the ways in which the body mediates itself to itself, for example through the 'nervous fluid' that was thought to flow through the nerves, which were therefore believed for centuries, against all the visible evidence, to be pipes made hollow for the purpose.

At the end of the nineteenth century a new connective carrier of the idea of carriage and connection came to be articulated. Galvanism suggested that the bodies of living creatures might be charged with a special kind of animal electricity. Alessandro Volta seemingly demonstrated conclusively that the electrical flows that characterised or operated on animal bodies were in fact not a special kind of vital force, indwelling in living creatures, but the same kind of electricity that operated in the world (Volta 1793). In fact, however, Volta's experiments opened up a huge repertoire of ways of imagining the operations of different kinds of fluid in the human body. In part, this was because electricity moved into the place occupied by an even more abstract kind of fluid, or rather a fluid that was paradoxically the idea that the very idea of fluidity itself could have an embodied form. For physicists, this meta-medium was the ether, at once that through which everything passed and that which could pass through everything else. For the expanding, itself ever more broadly mediated field of medicine, this meta-medium, where it was not identified with magnetic fluid, was regularly seen as some kind of electricity. Once electricity could be made to flow in a continuous current, as Volta's battery made possible, rather than in the jumps and jolts of electric shocks, it became conceivable for it both to do mechanical work, and to be a medium of communication, for now the flow could be both artificially maintained at a given level and also broken up at will, for example to create a coded series of dots and dashes, as in the telegraph. It also became possible for it to do the more abstract work of symbolising the idea of bodily flow, an abstract work that was always also practical and embodied. Not surprisingly, early electrical therapies were thought to be particularly effective in removing obstructions, or restoring functions like the menses when they seemed to have been blocked (Bertucci 2016: 124-5).

Electricity became identified with the vital principle, allowing for a physical solution to the philosophical problem of reconciling the idea of life on the one hand as an all-or-nothing quality (once you were dead, you were wholly and irreversibly dead, not just by and large so) and the irresistible tendency on the other to see life in quantitative terms, that is to say, as capable of flaring and flagging, or being more or less intensely present in different circumstances. Not surprisingly, the first experiments, physical and conceptual, with the idea of electrical life force concerned the possibility of restoring life to dead matter. One of these was the attempt in 1819 by James Cumming, Professor of Chemistry at the University of Cambridge, to restore to life the body of an executed criminal by means of a galvanic battery (Morus 2011: 11-19).

In fact, the huge advantage of electricity over any other candidate for Newton's medium of media, the subtle fluid that filled all space, was that it was not only universally inferrable, but could also be made visible and palpable. The commonest demonstrations of electrical force illustrated just this capacity for it to mediate between the demonstrable and the deducible. Most of the spectacular effects obtained from electrical displays during the eighteenth century depended upon the demonstration that human beings could act as batteries. If they were insulated, or, like the famous Charterhouse charity-boys employed by Stephen Gray, suspended by silk in mid-air, they could be charged up with static electricity, which they would not feel and of which they would give no sign until the electrical force was discharged by a conductor (Heilbron 1979: 247). It was as though electricity could allow you to feel your own faculty of feeling.

Another signal advantage of electricity as a therapeutic principle was that it conjoined form and force. It was easy to imagine the ways in which electric current might counter different kinds of congestion, impediment, or debility (paralysis, constipation, melancholia, hysterical aphonia) by imparting vitalising motion. Here, it was as though the electricity were simply spreading healing life and warmth into localised areas of lifelessness. But electricity also required specific kinds of ordered material arrangement, of conductors, resistors, batteries, contacts, and cables, in order that its flows could be channelled and regulated, which suggested, not just the power of vitalising, but also the power of restoring shape, system, and regularity to bodies that had become disordered in various ways (indigestion, mania, *tic douloureux*, epilepsy). Electricity could animate, that is, but it could also articulate. Its common association with other candidate forms of vital force was that it could be imagined both as a substance, which could be collected and stored in batteries as well as occasionally visualised, and as an energy, the nature of which was always to be, or wish to be, in motion. Not only was electricity the motive principle of bodily life, it seemed under certain circumstances capable of having its own kind of body. In fact, some early speculators suggested that the 'subtle aether' of electricity might itself come in different forms in the body, some more lively and mobile than others, the usefulness of fresh electricity being to force stagnant forms out, like water flushing a blocked drain:

> it may not be the safest Way to trust to one or two Shocks only, for that may, perhaps, be only removing the stagnant Aether from one Part of the Body to another, which Method, in all Probability, cannot be so eligible, nor the Aether so salutary and enlivening, as if moderate Shocks were oftener repeated, and the

> active Spirit, or electrical Aether, were sure to move, not only through the Part affected, but also that, by this Means, such stagnant Aether may be displac'd and driven quite out of the Body by the new. (Lovett 1756: 136)

Early in the nineteenth century, when electrical intervention was still reliant on the application of local shocks, the same discourse of carefully-judged modulation is in evidence. According to the accounts given in John Birch's *Essay on the Medical Application of Electricity* (1803), a patient suffering pain from inflammation of the knee who is reluctant to expose himself to electrical shock is reassured to find that electricity in fact eases his sensitivity; a servant girl is given violent but relieving laxation from the application of electricity; and the use of electric shock is commonly reported to restore muscular tone and sensation in the case of paralysis following stroke, for example in the case of a 60-year-old man who was 'blighted' by exposure to 'a bleak wind', such that 'his eye-lid dropped, his tongue hung slabbering from his mouth, his face was drawn on one side' (Birch 1803: 21-2). So improved was he after application of electricity that he returned every year in the Spring 'to have a course of electric operations, which he thought *reanimated* his nerves' (22). There were also more drastic applications, as in the case of a fishmonger who presented with an acute headache accompanied by visual disturbance:

> Confiding in the notion of the electric shock being harmless, under proper management, I did not hesitate to pass one through the brain [...] He was instantly relieved from his acute pain, the sight of his eye became perfect, and he thought himself cured as by magic. (25)

A powerful shock delivered to a tumour of the testicle succeeded in dissolving the tumour where all other treatments ('mercurial friction [...] vomits, cataplasms, and other external applications', 35) had failed.

One of the most important things about electricity when it came to investigating and imagining its medical applications was the fact that it was so conspicuously dangerous. As an extension of the principle that what does not kill you will make you strong, we tend to believe that only things capable of killing you can do you any real good. With the development of the Leyden jar, which did not require much in the way of specialised equipment but could generate very large electrical charges and deliver equivalently large electric shocks, it became perfectly clear that electricity was capable of causing pain, injury and even death. This made for the kind of complex adjustment and management of levels and quantities – the pseudo-precision of the exact dose, a word adopted into English from Latin *dosis*, something given, at the beginning of the seventeenth century – that I have noted as an important part of the symbolic efficacy of medical procedures.

As a powerful force, electricity partook of the ambivalent logic of the *pharmakon*, the poison that could, if managed correctly, be a remedy. Electricity was a force that it was both necessary to manage and that was capable of being managed. 'Electricity, Gentlemen' warned Herbert Tibbits, in a lecture given at the West End Hospital, 'is by no means one of those remedies that, failing to do good, is unlikely to do harm. On the contrary, in injudicious hands, it is potent for evil, while the benefit to be derived from it is in exact proportion to the judgment and care with which it is administered' (Tibbits 1879: 3). The authority of the medical practitioner, along with the patient's willingness to believe in the efficacy of the treatment, are both considerably enhanced by the sense of danger or damage held expertly at bay. Tibbits writes, for example, of the fearsome-sounding Faradaic 'wire brush' for applying electricity to the skin, that

> [w]ith a strong Faradaic current this wire brush becomes the most powerful of all the excitants of the skin which do not disorganize its structure; in fact it was proposed by some scientific parliamentary philanthropist as a substitute for flogging in the navy, and I have no doubt that more intense pain may be produced by it than by any application of the cat, however well laid on! (41)

The need to take care with the application of the medical procedure becomes an enactment of the care being taken with the patient's body. Imaginary machineries turn out to be much more effective, or at least much more convincing, if they imagine constraints and difficulties for themselves rather than simply sweeping past them into wish-fulfilment. In order to work, medical machines need to appear to do work, and need to appear to need work.

In the early days of electrical therapeutics, James Graham, whom we have already encountered in Chapter Four as a purveyor of sexual pleasures, ran a kind of high-class medical bordello he called the Temple of Health, where visitors could sleep on a specially-designed Celestial Electrical Bed which he promised would guarantee conception. Graham commissioned a talented mechanic called Thomas Denton to construct the bed, whose many intriguing moving parts included automata and mechanical music-making (Syson 2008: 181-6). But he went to considerable trouble in his verbose advertisements to play down the mechanical nature of the experience, representing the electricity as much as possible as something organic:

> For his own part, he has been taught by reason, and by the most attentive observations in innumerable experiments on almost every substance in Nature, and in a course of practice in the cure of diseases, far more extensive than that of perhaps any other man in the world, he has been taught, that all violence is

hurtful; that ELECTRICAL SHOCKS OUGHT SELDOM OR NEVER TO BE GIVEN; that partial frictions, sparks, and brushings with rich medicinal substances charged with electricity, or gently pervading the whole system with a copious tide of that celestial fire, fully impregnated with the purest, most subtle, and balmiest parts of medicines, which are extracted by, and flow softly into the blood and nervous system, with the electrical fluid, or restorative aethereal essences. In those cases where shocks are absolutely necessary, instead of charging bottles lined with tin foil, or other gross, impure, and perhaps arsenical metals, his jars are filled with loadstones, sulphur, quick-silver, and with the mildest, yet most active medicinal substances from the animal, mineral, or vegetable kingdoms, including [...] 'the powerful and salutary *Effluvia* of antimony, aromatic oils, Peruvian bark, castor, camphire, musk, ambergrease, the balsams of juperin, Peru, Tolu and Gilead, and the influences of electricity, air, aethereal medicines, and magnetism. (Graham 1778: 26-7)

Graham did everything he could to represent electricity as a gaseous phenomenon, emphasising the delicious fragrancy of his ethereal medicines and plasters and combining electricity with 'fixed air' (carbon dioxide) to provide effluvial vapours and baths.

The development of electrotherapy occurred at the same time as a number of other developments. There was, first of all, the development of electrical engineering itself. For most of the century this was focussed on telegraphy, since the uses of electrical apparatus for lighting, heating, transport, and domestic uses would have to wait for the development of large power grids. There was however a remarkable rise and multiplication of equipment of various kinds, much of it powered by steam or combustible gas. Much of the resulting equipment – presses, pumps, mills, locomotives – operated on scales larger than that of the individual body, and was industrial rather than domestic. This technological context interacted with political, social, and commercial factors. Medicine was increasingly professionalised and centred on the institution of the hospital; but, in tension with this was the growth of communication and advertising, with the possibility it seemed to open up of a market in medical provision, along with the possibility of chicanery and charlatanism. Reputations and fortunes were made and lost from imaginary electrical devices, one of the most conspicuous of these fraudulent outfits being the Medical Battery Company run by Cornelius Bennett Harness. This company marketed many different appliances from the 1880s onwards, but in particular an electropathic belt, which it was claimed would be sovereign against headache, palpitations, indigestion, sleeplessness, 'brain fag', debility, and a host of other rather notional disorders. Harness's downfall began in 1892, when he was sued by a dissatisfied customer, triggering a series of similar claims and a number of enquiries which publically demonstrated the fraudulent nature of his claims (Loeb 1999).

It would be easy to see medical electrotherapeutics entirely as an expression of Victorian credulity, quirkiness, and quackery. There is no doubt that it was often mocked in these terms by contemporaries. One of the most prominent of medical hoaxes in the late nineteenth century were Elisha Perkins's Metallic Tractors: pointed metal rods, which Perkins claimed would allow him to draw away pain and disease simply through passing them over the body of the sufferer. Perkins was expelled by the Connecticut Medical Society and his contrivance universally derided, most particularly in a comic poem by Thomas Green Fessenden of 1804 which was still being reprinted as late as 1837. Fessenden jeeringly associated the Metallic Tractors with galvinism and the suggestion that it might raise the dead:

> With powers of these Metallic Tractors,
> He can revive dead malefactors;
> And is reanimating daily,
> Rogues that were hung *once*, at Old Bailey!
>
> And sure I am he'll break the peace,
> Unless secur'd by our police;
> For such a chap, as you're alive
> Full many a felon will revive.
>
> And as he can (no doubt of that)
> Give rogues the *nine* lives of a cat;
> Why then, to expiate their crimes,
> These rogues must all be hung *nine* times. (Fessenden 1803: 63-5)

The temptation is to read this history in terms of the struggle between real apparatus, developed through experimental means in professional contexts, and fantasy apparatus marketed by the unscrupulous to the witless or desperate. There certainly seems to have been concern among some medical practitioners at the fact that electrical apparatus was being marketed for people to use on themselves without medical supervision. And yet the relative cheapness and easiness of operation of electrical equipment meant that the possibility of self-administered electricity was there from the beginning, and may just as often have bound doctor and patient together in rituals of accredited performance as put them at odds with each other. John Reddall, a popular lecturer on electricity, wrote in 1760 that it would be 'worth while, for any paralytic person, who can afford it, and has conveniency, to have an electrical machine in their house' (quoted in Lovett 1760: 40), and John Wesley was among those who encouraged ordinary people to furnish themselves with cheap electrical machinery for therapeutic self-administration (Bertucci 2006: 35). A 24-year old patient known as C.B. who had suffered for ten years from 'the whole train of hysteric symptoms', including cramps in all parts of her body, convulsions and a 'choaking *deliquium*' [fainting] (Evans 1776: 84), all thought to have been triggered by 'an obstruction of the *menses*, from imprudently exposing herself to cold, at the time of their appearance' (83), described her treatment in Philadelphia by Benjamin Franklin:

> I receiv'd four strokes morning and evening; they were what they call 200 strokes of the wheel, which fills an eight gallon bottle, and indeed they were very severe [...] I staid in town but two weeks, and when I went home, B. Franklin was so good as to supply me with a globe and a bottle, to electrify myself every day for three months. The fits were soon carried off. (85)

Patient, practitioner, supplier, and apparatus all participate in the formation of an imaginary apparatus and apparatus of imagining of the body electric, which encompasses not only the fantasy of a bodily actuality but also the actuality of a widespread fantasy. There is a story told of a visitor to the physicist Niels Bohr who was surprised to see a horseshoe hung over a door for luck. Surely a man of science could not attach any credence to such things? Naturally not, Bohr replied, but 'I've sometimes noticed that it works even when you don't believe in it' ('The Tagarene Shop' 1956: 422). All one needs, in such cases, it seems, is the belief in other people's belief, or even the belief in other people's belief that yet other people think there might be something in it. The fact of being imaginary is not in itself any great disadvantage to an imaginary machinery. It is in this sense that an imaginary machinery of the body electric might have been developed and maintained alongside the doubts or even outright mockery of experts. Perhaps the most convincing demonstration of the dominance of the electrical idea of the body is, oddly enough, precisely the fact that it seems so abruptly to have disappeared from view after around the end of World War I (Gilman 2008: 339). All of a sudden, a set of procedures that were part of a shared texture of experience and expectations dissolved, revealing how much part of the field of assumed possibility they had been.

As the nineteenth century wore on and electrical procedures and machinery became more established in hospitals, while electrical engineering developed in subtlety and reach, medical therapeutics began to be subject to more precise and systematic kinds of measurement. This outlook displaced the fantasy of electricity as a special kind of life force, with the body as a favoured locale for its operations. Instead, as Iwan Rhys Morus has proposed, the body was integrated into larger and more various electrical networks. If 'the body was coming to be surrounded by instruments and practices that allowed it to be reconfigured as an electrical machine' (Morus 1999: 271), the effect was to decathect this machine, reducing the voltage of the imaginary investments in it. As the body became connected up in ever more ways to grids and networks, the idea of the body electric lost its cohering power. The electrified body – not least in the adoption of electrocution as a means of execution in the USA – took the place of the body conceived of as in its essence and functioning as electrical. If indeed '[t]he human body could be imagined as a section of the Atlantic cable, an electrolytic conductor, an induction coil or any number of other electrical devices [and] […] treated as if it were a component of machinery and approached with the practices appropriate to dealing with machines' (278), this meant that it had in effect been dissolved into a series of mechanical components rather than concentrated into a phantasmal body machine. Imaginary machines need to maintain a balance between the palpable and the potential. One continues to have to imagine pretty much everything about electricity, but electrical devices perhaps became too familiar for the force of belief in the special mysterious power of electricity to be maintained.

Another of the advantages of electricity to the medical imagination, along with its capacity to suggest the possibility of managing and manipulating various kinds of magical fluid, was its association with action-at-a-distance, as explicated in the principles of electromagnetic induction, discovered by Michael Faraday in 1831. In fact, the associations between electricity and induction had been observed, if not well understood, since the eighteenth century. Electricity was not just an art of exquisite and subtle contacts, but also one of what was called 'influence'. In fact, before electricity had been made to flow in currents it existed as clouds or fields of charge that were able to exercise forms of attraction and repulsion at a distance. There is a tension in every machine between what it is and what it does, between form and force. This tension had always been particularly marked in the case of electricity, given the relative ordinariness of the materials required for the apparatus (glass, tinfoil, water, copper) and the capacity of the generated force violently to outrun its mechanical beginnings. Tension was in fact a word that was taken up into the electrical field, with George Adams writing in 1785 that '[[t]he whole energy of electricity depends on its tension, or the force with which it endeavours to fly off from the electrified body' (Adams 1785: 208). From the 1890s onwards, the allure and authority of electricity in medical machines gave way to machines centred on the powers of radiation, in which the tension between force and form was at its greatest. These new machines to imagine with form the subject of the next chapter.

6

Radiation Machines

The material and the mechanical are so closely twinned as to seem synonymous. A mechanist view of the world is one that does not allow for any kind of metaphysical entities or principles. The world of matter, on this view, is held to be fully explicable by laws that are mechanical in their regularity and their totality. Mechanism is simply how matter works, in the absence of anything else but matter. And yet, we have seen that, in their need to be imagined, machines can seem to involve the occult, the invisible, the anoptic (that which is not of the order of the visible), or the imaginary, so that they can seem to exceed or escape from the conditions of self-evidence. Electromagnetic machines in particular seemed to suggest the paradoxical possibility of mechanical actions performed without the mediation of any kind of physical apparatus. These may look forward to the conditions of what, paradoxically enough it might seem, came in the 1920s to be called 'quantum mechanics', a mechanics that was calculable yet not in any easy sense imaginable – or perhaps we might just as well say, could only be imagined as that which did not yield easily to being imaged. Attempts to subject quantum mechanics to ordinary ways of conceiving mechanical relationships need to have recourse not to mechanical metaphors, of which there is always such an abundance, but new metaphors for mechanism itself – like, for example, the modified noughts-and-crosses game of 'Quantum Tic-Tac-Toe' proposed by Alan Goff as a way of visualising quantum-mechanical processes (Goff 2006).

Quantum mechanics was preceded in this by radiation theories. Radiation was the successor to electricity as the bearer of a kind of phantasmal mechanics. In a series of popular and successful novels following the appearance of her novel *A Romance of Two Worlds* (1886), Marie Corelli had developed a supernaturalist system centred on the spiritualised powers of electricity. In this Corelli echoed the envious hunger for scientific validation that was characteristic of many varieties of late Victorian supernaturalism. Like Helena Petrovna Blavatsky, the originator and high priestess of theosophy, Corelli was vehemently opposed to what she regarded as the vulgar theatrics of spiritualism, and attempted to give her theories of spiritual evolution a rational basis by assimilating ideas derived from contemporary science and technology. Corelli was less religiously eclectic than Blavatsky, and continued to make Christianity the centre of her 'Electric Creed', in which God becomes 'a Shape of Pure Electric Radiance' (Corelli 1886: 2.121), and Christ an '*electric flame* or *germ* of spiritual existence combined with its companion working force of *Will-power*' (2.124). Modern science demonstrated that the miracles of Christ were electrical phenomena: walking on water, she coolly affirmed, is 'a purely electric effort, and *can be accomplished now by anyone* who has cultivated sufficient inner force' (2.135-6).

Twenty-five years on, in the Prologue to her novel *The Everlasting Life*, she announced her conversion from the idea of electricity as a vital principle to a more contemporary analogy to 'the infinite power of *that* within us which we call Soul, – but which we may perhaps in these scientific days term an eternal radio-activity, – capable of exhaustless energy and of readjustment to varying conditions' (Corelli 1911: 14). Indeed, she claimed that she had been in on the secret even when writing her debut novel:

> I was forbidden, for example, to write of *radium*, that wonderful 'discovery' of the immediate hour, though it was then, and had been for a long period, perfectly well known to my instructors, who possessed all the means of extracting it from substances as yet undreamed of by latter-day scientists. I was only permitted to hint at it under the guise of the word 'Electricity' – which, after all, was not so much of a misnomer, seeing that electric force displays itself in countless millions of forms. (18)

Corelli's heaving blend of overheated romance, vitalist metaphysics, and occultism, with plentiful hints of clairvoyance, reincarnation, mesmerism, Egyptian mysticism, and mysterious psychic powers and traditions, had secured her position as one of the most popular and successful authors of the Edwardian period, outselling writers like Wells, Barrie, Conan Doyle, and Kipling. Both Corelli and her opinionated and high-minded heroines are contemptuous of the atheistic scientific mechanism of their day: in 1910, she wrote a Gothic short story entitled *The Devil's Motor*, which emphasises the horrors of acquisitive materialism by putting the devil behind the wheel of a car that poisons and crushes everything in its path:

> In the dead midnight, at that supreme moment when the Hours that are past slip away from the grasp of the Hours yet to be, there came rushing between Earth and Heaven the sound of giant wheels, – the glare of great lights, – the stench and the muffled roar of a great Car, tearing at full speed along the pale line dividing the Darkness from the Dawn. (Corelli 1910: n.p.)

Nothing much happens in the story, except for the evocation of Blakean images of industrial chimneys 'belching forth sickening smoke' amid 'deafening noise and crashing confusion as of ten million hammers beating incessantly', with the Fiend exultantly celebrating the descent of the modern world into futility and death, as he careers his infernal vehicle at breakneck speed into the dark abyss signifying the end of the world:

> 'you shall bend the lightning to your service, and the lightning shall slay! [...] you shall seek to bind the winds, and sail the skies, and Death shall wait for you in the clouds and exult in your downfall! Come tie your pygmy chariots to the sun, and so be drawn into its flaming vortex of perdition!' (n.p.)

Although she did not adhere to any particular supernaturalist doxa, Corelli's willingness to blend the psychic and the pseudo-scientific along with her attraction to the idea of hierarchies of the spirit gave her work strong affinities with theosophy. *The Life Everlasting* is split between its denunciations of the spiritually barren mechanism of modern life and suggestions that the magical powers of technology might not be wholly inimical to spiritual progress. Her stories are in fact thronged with mysterious psychically-powered devices and machineries, making it apt that, when the 'telly' arrived later in the century, it was referred to in Cockney rhyming slang as a 'Marie'. The central relationship of *The Life Everlasting* is between a free-thinking 'psychist' young woman, whose name we never discover, and Rafael Santoris, the mysterious man she meets during a yachting trip in Scotland, who is the captain of 'The Dream', a spectral vessel powered by some photosynthetic version of electricity that allows for light to be turned into kinetic force. Santoris is a mechanical Magus as well as a mystic:

> 'Our yacht's motive power seems complex, but in reality it is very simple, – and the same force which propels this light vessel would propel the biggest liner afloat [...] even when wider disclosures of science are being made to us every day, we still bar knowledge by obstinacy, and remain in ignorance rather than learn. A few grains in weight of hydrogen have power enough to raise a million tons to a height of more than three hundred feet, – and if we could only find a way to liberate economically and with discretion the various forces which Spirit and Matter contain, we might change the whole occupation of man and make of him less a labourer than thinker, less mortal than angel! The wildest fairy-tales might come true, and earth be transformed into a paradise! And as for motive power, in a thimbleful of concentrated fuel we might take the largest ship across the widest ocean. I say if we could only find a way! Some think they are finding it – ' (Corelli 1911: 151-2)

Using this mysterious spiritual propulsion, 'The Dream' is not only able to fill its sails when there is no wind (though it is never explained why it should in that case need sails at all), but even to power its progress with moonlight, rather like the novel it inhabits. The novel indeed pulses and glitters relentlessly with various kinds of radiance and radiation, white, pearly, warm, lustrous, 'weirdly gleaming' (359), 'soft yet effulgent radiance [...] as if the walls glowed with some surface luminance' (323-4).

The theory of human aura, popularised by writers like the theosophist Charles Leadbeater, makes its predictable appearance, in the exposition given by the all-knowing Santoris of how he knows his old college friend Harland is ill:

> A positive 'light' surrounds you – it is exhaled and produced by your physical and moral being, – and those among us who have cultivated their inner organs of vision see *it* before they see *you*. It can be of the purest radiance, – equally it can be a mere nebulous film, – but whatever the moral and physical condition of the man or woman concerned it is always shown in the *aura* which each separate individual expresses for himself or herself. (159)

Late in the novel, its heroine, launched on a path of spiritual instruction, encounters the central principle of magical thinking, in a 'magic book' entitled *The Secret of Life*. Her quotations from the book accord neatly with Freud's characterisation of magical thinking as 'Allmacht der Gedanken', 'omnipotence of thoughts' (Freud 1991: 9.105; Freud 1953-74: 13.84), though mediated through contemporary technologies:

> 'Thought is an actual motive Force, more powerful than any other motive force in the world. It is not the mere pulsation in a particular set of brain cells, destined to pass away into nothingness when the pulsation has ceased. Thought is the voice of the Soul. Just as the human voice is transmitted through distance on the telephone wires, so is the Soul's voice carried through the radiant fibres connected with the nerves to the brain. The brain receives it, but cannot keep it – for it again is transmitted by its own electric power to other brains, – and you can no more keep a thought to yourself than you can hold a monopoly in the sunshine.' (Corelli 1911: 373-4)

Neatly and sweetly, the magic book skewers itself on the paradox that attends every evocation of universal and unlimited mediation – namely that it would not have any effect at all unless it were in some way limited:

> 'Everywhere in all worlds, throughout the whole cosmos, Souls are speaking through the material medium of the brain, – souls that may not inhabit this world at all, but that may be as far away from us as the last star visible to the strongest telescope. The

harmonies that suggest themselves to the musician here to-day may have fallen from Sirius or Jupiter, striking on his earthly brain with a spiritual sweetness from worlds unknown, – the poet writes what he scarcely realises, obeying the inspiration of his dreams, – and we are all, at our best, but mediums for conveying thought, first receiving it from other spheres to ourselves, and then transmitting it from ourselves to others. Shakespeare, the chief poet and prophet of the world, has written: "There is nothing good or bad but thinking makes it so," – thus giving out a profound truth, – one of the most profound truths of the Psychic Creed. For what we *think*, we *are*; and our thoughts resolve themselves into our actions.' (374)

The magical credo, 'what we *think* we *are*', has a positive and a negative aspect. On the positive side, it promises a world subdued to will. On the negative side, it threatens the possibility of becoming a captive to one's own thought. If the world is to be subdued to thought, then thought must itself be subdued to will; but that is an unwinnable struggle if 'you can no more keep a thought to yourself than you can hold a monopoly in the sunshine' (374). If your thought can penetrate and control everything, then it can also penetrate you, leaving you merely transparent, the will-less vehicle of thought, spilling in all directions, rather like radiation. The fantasy of being in touch with a universal power, extending indifferently in all directions and with no modulation of its force. makes an absurdity of the other side of the fantasy, that this knowledge might actually have, in the mechanical sense, an Archimedean point, or point of application, any way of acting in one way rather than another, as *The Secret of Life* promises:

'If you would stand firm, you must stand *within* the whirlwind; if you would maintain the centre-poise of your Soul, you must preserve the balance of movement, – the radiant and deathless atoms whereof your Body and Spirit are composed must be under steady control and complete organisation like a well disciplined army, otherwise the disintegrating forces set up by the malign influences of others around you will not only attack your happiness, but your health, break down your strength and murder your peace.' (392)

The pan-psychist immediation that makes *thought* and what the spiritualist journalist W.T. Stead called *throughth* (Stead 1893: 426) indistinguishable also makes it impossible for there to be any distinguishable thought bearing on anything in particular, or any particular or determinate action of thought possible, thus busting apart the Psychic Creed and much else besides. Included in this much else must, of course, be the whole machinery of narrative that has brought the book to this point.

Indeed, narrative seems to have been entirely abandoned by this point in the novel, the final third of which is given over to a description of the narrator's spiritual ordeals in the House of Aselzion, located on the northern coast of Spain. The House of Aselzion is full of fantasy apparatus, mostly embodied in forms of magical architecture. There is automatic mood-lighting, her room illumined by a 'soft yet effulgent radiance' (Corelli 1911: 323) seemingly supplied by no lamps or burners, but 'as if the walls glowed with some surface luminance' (323-4); doors that open automatically or walls that melt away, baths that fill themselves; a dumb waiter arrangement whereby the dresser in her room is furnished with meals which are then automatically carried away; 'a crystal globe which appeared to be full of some strange volatile fluid, clear in itself, but intersected with endless floating brilliant dots and lines' (349), which makes visible the Brownian dance of atoms that are not subjected to the force of Will, and is intended 'as an object lesson, to prove that such things *are* – they are facts, not dreams' (350); and Aselzion's power of projecting images of the narrator's desire on to the surface of the sea, along with mysterious sourceless voices, rooms that appear and disappear, walls and staircases that melt away and mysterious cowled figures who abruptly appear and disappear. Of course, these are not machines, precisely, since their mechanisms are not open to view, and they work without any obvious workings. But they are not machines precisely because being not-machines is what they are there for – that is, to be imaginary machines, machines imagined because unimageable, imagined in order to be unimageable, machineries of phantasmal self-sublimation that manifest the melting away of machinery. As such they are avatars or outworks of the most powerful machinery of all, that of fantasy: or the thing that makes fantasy actual and able to work on the world, writing.

Most importantly for the novel, there are different forms of powerful rays, like the gaze of Aselzion, who 'studied my face with a keen scrutiny which I could *feel* as though it were a searching ray, burning into every nook and cranny of my heart and soul' (311). Rays are not just instruments of inspection, but are also often the means of transmitting ideas, voices and images. The altar of the chapel is adorned with a large cross which gives out

> ever darting rays of fiery brilliancy, and the effect of its perpetual sparkle of lambent fire was is if an electric current were giving off messages which no mortal skill would ever be able to decipher or put into words but which found their way into one's deepest inward consciousness. (329)

The danger increases as the narrator enters the chapel on her own and approaches the central power source, which voluptuously probes, penetrates and permeates her in a mixture of mystical rapture and alien abduction fantasy: 'the glowing radiance of the Cross and Star in all that stillness was almost terrible! – the long bright rays were like tongues of fire mutely expressing unutterable things!' (341). Despite, or because of her terror, she is irresistibly drawn towards the 'perfect vortex of light […] that strange starry centre of living luminance' (342), until she is swallowed up – if she does not herself do the swallowing – in ecstatic consummation:

> step by step I went on resolutely till I suddenly felt myself caught as it were in a wheel of fire! Round and round me it whirled, – darting points of radiance as sharp as spears which seemed to enter my body and stab it through and through – I struggled for breath and tried to draw back, – impossible! I was tangled up in a net of endless light-vibrations which, though they gave forth no heat, yet quivered through my whole being with searching intensity as though bent on probing to the very centre of my soul! (342)

Corelli's work is distinctive for the ways in which it allows infantile delusions of omnipotence to be articulated alongside the delicious fantasy of being invaded by external powers. As Aselzion explains to her, the narrator must be subjected to the mechanics of influence in order herself to become the wielder of omnipotent thought:

> 'When I took you up to your room in the turret, I placed you under my influence and under the influence of four other brains acting in conjunction with myself. We took entire possession of your mentality, and made it as far as possible like a blank slate, on which we wrote what we chose. The test was to see whether your Soul, which is the actual You, could withstand and overcome our suggestions. At first hearing, this sounds as if we had played a trick upon you for our own entertainment – but it is not so, – it is merely an application of the most powerful lesson in life – namely, the resistance and conquest of the influences of others, which are the most disturbing and weakening forces we have to contend with.' (411)

The self-possessing soul is the one who can stand up against 'disturbing or opposing brain-waves' (410), ' "influences" – the magnetic currents of other brains' (412), influences to which a woman is particularly subject, allowing herself to be subdued by the wishes of a man and thereby sinking into the condition of 'a mere domestic drudge or machine' (412). But it also appears as though Corelli were engaged a struggle for imaginative possession of her own fantasy of influence and its effect on her readership:

> 'Remember,' he said – 'what force there is in a storm of opinion! The fiercest gale that ever blew down strong trees and made havoc of men's dwellings is a mere whisper compared with the fury of human minds set to destroy one heaven-aspiring soul!' (415)

We learn that the only real thing amidst all of the phantasmal objects and contrivances the narrator has experienced is *The Secret of Life*, the book-within-a-book that teaches the essential wish-fulfilling and reality-discounting lesson that ' "All action is the material result of Thought" ' (413).

The mechanics of the narrative are to be elided with the imaginary mechanisms in the narrative: they are there to give evidence of a power that commands and compels the material world, but precisely in order to suggest a power to go beyond it. This power is in fact to be identified wholly with the fantasy of the writer, able through sheer force of will to summon and determine everything in her narrative world. Everything depends, as in the Gothic elaborations of sadomasochistic fantasy, on the machinery that, seeming to constrain the subject of fantasy, in fact expresses its limitless power, even and especially over its own desire for dominion. Aselzion doubts that the narrator has the strength to undergo the ordeal that her lover has previously endured. The narrator has to persuade the monkish master-adept Aselzion that she is capable of undergoing the process of spiritual instruction that has given her lover his material and spiritual power. Her voluptuously imperious obedience is the model for that of every Gothic heroine determined to win victory through suffering (since the imperious Master is always the puppet of her fantasy, every humiliation he inflicts will be further proof of his helplessness).

> 'As I have just said,' he went on – 'this is no place for women. The mere idea that you should imagine yourself capable of submitting to the ordeal of a student here is, on the face of it, incredible. Only for Rafel's sake have I consented to see you and explain to you how impossible it is that you should remain – '
> I interrupted him.
> 'I must remain!' I said, firmly. 'Do with me whatever you like – put me in a cell and keep me a prisoner, – give me any hardship to endure and I will endure it – but do not turn me away without teaching me something of your peace and power – the peace and power which Rafel possesses, and which I too must possess if I would help him and be all in all to him –' (314)

It may seem as though Marie Corelli has left the vulgarity of science-fiction mechanics behind, in substituting metaphysics for narrative. Indeed, this seems to be the point of the absurd and pompous Author's Prologue to *The*

Life Everlasting, in which Corelli reviews her career, paying off old scores, ridiculing (ridiculously) her ridiculers, glorying with inglorious immodesty in her massive celebrity and success, while yet hinting that most of the important content of her novels is in fact above the heads of all but a select few – even, at one point, tearing a strip off her imaginary orthodox reader:

> I know very well, of course, that I must not expect your appreciation, or even your attention, in matters purely spiritual. The world is too much with you, and you become obstinate of opinion and rooted in prejudice. Nevertheless, as I said before, this is not my concern. Your moods are not mine, and with your prejudices I have nothing to do. (12)

It is as though Corelli were raging against the necessity of having to write for such readers at all, all the time hinting that there are much higher, more exclusive purposes at work than shifting barrow-loads of psycho-Gothic fantasy aids. But in fact it is precisely this will to transcendence that represents the triumph of the imaginary mechanism of imagining itself. The various machines in the narrative, actual and metaphorical, are metaphors for the work of narrative, which this narrative is working to undo – but they turn out to be metaphorical enactments of the dependence of all narratives on the mechanisms of mediation. This is in fact the inversion of the inversion of sadomasochistic fantasy; where the whole point of a sadomasochistic scenario is to ensure that every snare and affliction becomes a triumph of the will that is capable of willing its own subjection, a subjection to a will that is always also in advance its own (and yet for that reason must neither be wholly known or governable, so must in fact be as resistant to persuasion as a machine). The fantasy of a fantasy that has absolute dominion is in fact the machine of fantasy in operation. It is a knitting machine that knits itself in and knits itself out, and Corelli struggles to avoid the recognition that, since it is engaged on making itself, such a machine is not sublimatable – that, in Conrad's grimly grinning words, 'you cannot by any special lubrication make embroidery with a knitting machine' (Conrad 1983: 425). In her Author's Prologue, Corelli quotes *Hamlet*: 'What a piece of work is man! – how noble in reason! – how infinite in faculty! – in form and moving how express and admirable! – in action how like an angel! – in apprehension how like a god!' (12). She pulls out for the reader's attention two of Hamlet's phrases:

> 'How infinite in faculty!' – and 'In apprehension how like a god!' The sentences are prophetic, like so many of Shakespeare's utterances. They foretell the true condition of the Soul of Man when it shall have discovered its capabilities. 'Infinite in faculty' –that is to say – Able to do all it shall WILL to do. There is no

end to this power, – no hindrance in either earth or heaven to its resolute working. (Corelli 1911: 12-13)

She keeps silence about the most well-known phrase in what she misquotes in the usual way, giving 'What a piece of work is man!' instead of the 1604-5 Quarto 'What piece of work is a man!' or the Folio 'What a piece of work is a man!' It is the working out of the work that it is the work of her prologue to conceal. Machination must cease, or be dissimulated in fantasy, a fantasy that must keep its mechanical nature under wraps.

In radiation effects and technologies Franz Reuleaux's definition of a machine as 'a combination of resistant bodies so arranged that by their means the mechanical forces of nature can be compelled to do work accompanied by certain determinate motions' (Reuleaux 1876: 35) had been superseded by machineries that seemed to operate without recognisable forms of limit, resistance, or impediment. Radioactivity was the ultimate phantasmal machine in being both mechanical and yet seemingly without material limit or containment. It was a machine that went beyond the conditions of being a machine.

> This is precisely what the radio-activity in each individual soul of each individual human being is ordained to do, – to absorb an 'unknown form of energy which it can render evident as heat and light.' Heat and Light are the composition of Life; – and the Life which this radio-activity of Soul generates *in* itself and *of* itself, can never die. (Corelli 1911: 19)

Corelli's novel belongs to an archive of influencing machines, to give them the name devised by Victor Tausk (1933), an archive that includes very much more than those rather exotic cases of 'technical delusion', the term employed within modern psychiatry, to which psychoanalysts and cultural historians tend to return: the cases of James Tilly Matthews, documented by his doctor James Haslam (Haslam 1988; Jay 2003); of John Perceval (Perceval 1962); of Friedrich Krauß, who documented his 'magnetic poisoning' in volumes that appeared in 1852 and 1867 (Krauß 1852, 1867, 1967); and, most famously, and most extensively analysed, of Daniel Paul Schreber, whose *Memoirs of My Nervous Illness* (2000), which first appeared in 1903, was analysed by Freud in an essay of 1911 (Freud 1953-74: 12.1-82). There was in fact quite an intimate relation between the idea of 'influence' and electrical science at the end of the nineteenth century, since 'influence machines' was the name commonly given to machines producing what is nowadays more usually known as electrostatic induction, first demonstrated by John Canton in 1753, whereby an insulated conductor brought into the vicinity of an electrically charged body develops a positive charge on one end and a negative charge on the other (Thompson 1888). Technical delusions of this kind seem to be distinctively modern, such adaptation of underlying psychological disorders and symptoms to accord with changing historical conditions being known as pathoplasticity (Bell et. al. 2005) – though we might wish to note that, whether gas, electric or laser-powered, influencing machines tend to exhibit a certain pathoplasticity on their own account, in their capacity to produce a bewildering variety of kinds of symptom and effect.

Louis A. Sass sees the schizophrenic's self-identification with machines as an expression of what he calls 'phantom concreteness, and of the reifying, distancing, often self-alienating processes that underlie it' (Sass 1994: 95). Thomas Fuchs agrees that 'An influencing-machine is […] the expression of a self-objectification' (Fuchs 2006: 35). But there is no good reason to think that machines can be equated to mere or pure objects, whether for psychotic patients or anyone else. On the contrary, there is always a difference between a machine and an object, if only in the fact that a machine is a communicative system of objects which therefore has a being for-itself as well as merely in-itself. Far from being an object, or a force of objectification, a machine that produces or exploits radiation seems always to embody a kind of charged condition, and thus a surrogate or enlarged life.

Marie Corelli was not alone in seeing radium as a pure source of perpetual energy, and therefore a kind of mechanics without limit. Marie Curie had been led to the discovery first of the element she named 'polonium' in honour of her native Poland, perhaps in ironic hope given how strongly determined it was by other powers, and then the element radium, by Becquerel's discovery that the X-rays emitted by uranium salts did not seem to depend on any external action but rather on some indwelling yet inexhaustible property of uranium itself. As Carolyn Thomas de la Peña suggests, this encouraged the idea that radioactivity was a principle of limitlessness; it was the fantasy gift to fantasy that never gave over giving:

> Radium entered into and flourished within a popular culture of energy fantasy well established by previous mechanical and electrical energy devices. Yet unlike machines' and electric devices' measurable limits and inorganic relationships to the body, radium was invisible, ingestible and seemingly infinite. Many believed it could power the body by a 'technology' as natural as the heart and muscles themselves. (Peña 2003: 174)

This capacity to generate and perpetuate itself from within meant that, as Luis A. Compos has shown, radiation, as embodied in the mutative element radium, was identified early on as more than just a new kind of physical force: as a new embodiment of the idea of Life itself, giving rise to a 'vitalised radioactive discourse' (Compos 2015: 20).

In one sense, radioactivity represented an entirely new way of conceiving the physical world, not in terms of objects and their interactions, but in terms of fields of influence. It is a world, not of stresses, resistances, contacts, and impacts, but of influences, emanations, permeations, participations, minglings, and superimpositions. Although radiation belonged to a new kind of physics, centred on fields rather than bodies delimited in space, there seemed to be considerable resources in poetic, religious, and magical tradition for imagining this kind of permeative geometry, and especially in relation to the idea of 'influence', a word which entered English from Latin *influxus stellarum* primarily to signify what the OED calls 'The supposed flowing or streaming from the stars or heavens of an etherial fluid acting upon the character and destiny of men, and affecting sublunary things generally'.

The history of the radioactive imaginary is one which sees largely positive and optimistic associations giving way to negative ones. X-rays started to be used to treat cancers almost immediately upon their discovery in 1896. The infatuation with the curative and invigorating qualities of radium has been amply and, it must be said, sometimes rather gleefully demonstrated. Nevertheless, the offensive and destructive powers of radiation had also long been apparent, and energetically dreamed of. The first report of the use of a radiant weapon is the story that Archimedes used a mirror to focus the light of the sun to create a burning glass that destroyed Roman ships during the Second Punic War (Wilk 2013: 12-20). It was perhaps the association of radiation with the effects of atomic warfare following Hiroshima and Nagasaki that produced the decisive change in the valence of radioactivity.

Rays embody the idea of an immaterial machinery of influence. A ray, from Latin *radius*, means a line, and was defined abstractly and arbitrarily by Newton at the beginning of his *Opticks* (1704) as 'The least Light or part of Light, which may be stopt alone without the rest of the Light, or propagated alone, or do or suffer any thing alone, which the rest of the Light doth not or suffers not' (Newton 1704: 1-2). The absolute abstraction of this notion means that a ray blends actuality and concept, suggesting that at its most reduced light might indeed consist of nothing but an idea of itself. A ray is often characterised as a reduced or perfected form of a beam of light, which is usually thought of as a broader column or pillar – hence the use of 'beaming' to mean broadcasting. This is assisted by the fact that rays and beams are both reversible, and were thought of, in the extromission theory of vision, as directed outwards from the eyes as well as inwards to them. The theory, first formulated by Empedocles, in which, according to David Park, '[t]he visual ray is like a long finger projecting from the eye, and sight is a kind of touch' (Park 1997: 35), survives in poetic and metaphorical uses, and may be thought of as an imaginary machine of vision. In such usage, the ray is thought of as intentionally produced or directed, and to be penetrative. X-rays would seem to represent a kind of mechanisation of the penetrative power often attributed to rays.

So we can distinguish rays from radiations: because rays are vectors, and can therefore be directed, they are light made intentional, or radiation in the indicative mood. The idea of the ray is an imaginary machinery of intended light. Ray-machines are externalised forms of a mechanical fantasy of sight. As such they can be thought of as weapons, and it is in this form that they make their appearance in fiction and fantasy from the 1890s onwards. The problem of how to direct and orientate radiation apparatus, or rather the conspicuous and triumphant deflection of that problem into fantasy, is central to one of the earliest Martian invasion narratives, Garrett P. Serviss's *Edison's Conquest of Mars* (1898), written as a serial for the *Boston Post* as a sequel to an unauthorised adaptation of Wells's *The War of the Worlds*, which had appeared under the title *Fighters from Mars* in the newspaper during January and February 1898 (Mollmann 2010: 388). The narrative begins in the period following an unsuccessful Martian invasion of Earth, which prompts a huge global effort to improve earthly technology to allow for a pre-emptive invasion of Mars. Central to this undertaking are two machines devised by Thomas Edison, making one of his many heroic appearances in fin-de-siècle fiction, here in the company of other living scientists like Wilhelm Roentgen, Sylvanus P. Thompson, and Lord Kelvin: an electrically powered flying machine which works by reversing the effects of gravity, and a disintegrator gun. The latter exploits what is described as 'the law of harmonic vibrations' applying to all matter in the universe (Serviss 1947: 13). Because every kind of matter is held to vibrate at its own characteristic frequency, all that is needed to destroy any particular portion of matter is some transmissible pulsation tuned to its frequency, on the model of the well-known belief that an army marching in step can cause a bridge to shake itself to pieces if it synchronises with its resonant frequency:

> Mr. Edison had been able to ascertain the vibratory swing of many well-known substances, and to produce, by means of the instrument which he had contrived, pulsations in the ether which were completely under his control, and which could be made long or short, quick or slow, at his will. He could run through the whole gamut from the slow vibrations of sound in air up to the four hundred and twenty-five millions of millions of vibrations per second of the ultra red rays. (14)

The extrapolating leap is from the idea of a universal principle of resonance to the idea that it might be susceptible to local direction and control, specifically in the form of a handgun that can project vibrations at the frequencies required to pick out and annihilate particular forms of matter, or even, as in Serviss's story, destroy particular portions of it. The idea of the ray-gun is perhaps the first fantasy of a medium of communication conceived as a weapon.

The issue is not that the idea of destructive mechanical resonance is without any foundation, but that it has crystallised into an autonomous fantasy. The fantasy survives in the widespread belief in the power of opera singers to shatter wine-glasses, of which there are few if any attested examples, as well as in the widespread will-to-believe (taking the form, as so often, of the will to believe in others' belief) in the 'Brown Note', the fear that very loud bass frequencies might cause involuntary collapse of the sphincter and evacuation. This latter fantasy is allied to suspicions that such effects are being investigated for military purposes, as embodied in Kate Bush's song and video *Experiment IV* (1986), which illustrates the search for 'a sound that could kill someone at a distance'. Kate Bush herself appears in the video first of all as a seductively wuthering siren and then as a banshee which escapes from the laboratory and lays waste to the entire research facility. The final shot, which has Kate Bush winking at the camera and putting her finger to her lips, teases the viewer with the delicious *folie de grandeur* that music might be the source of a kind of radioactive holocaust of pleasure.

Such acoustic machines have featured in *Fortean Times*. Sometimes the effects are morally rather than physically disintegrative, as for example in the Feraliminal Lycanthropizer, a fictional machine described in a pamphlet produced in 1990 by the American writer and conductor David Woodward, who also constructed versions of Brion Gysin's Dreamachine. The Lycanthropizer is a paranoid inversion of the Dreamachine, in that its effects are allied to Thanatos rather than Eros. Woodward describes the machine as 'a low frequency thanato-auric wave generator' that is '[k]nown for its use by the Nazis and for its animalizing effects on human subjects tested within measurable vibratory proximity'. The machine creates violence and sexual desire, its essential function being 'to trigger states of urgency and fearlessness and to disarmour the intimate charms of the violent child within [...] [T]he Trithemean incantations richly pervading the machine's aural output produce feelings of aboveness and unbridled openness' (Woodward 1990: n.p.). (The 'disarmour' reference suggests both military applications and a family resemblance to the Reichian orgone accumulator.)

The problem for every resonant weapon, whether of mass or minor destruction, is the same as that which had to be overcome by early theorists and engineers of radio broadcasting, namely how to tune a transmission in such a way as to prevent it going everywhere and being heard by everyone. It is the tuning principle which needed to be mastered in order for radio to become a means of communication rather than of general broadcast. The idea of tuning offered an important element of seeming objectivity to imaginary machines. For tuning depended, not on any organic form of affinity or sympathy involving the communication between qualities, but on quantitative adjustments, of frequencies and wavelengths. Such adjustments required the operation of care and careful operation, accompanied all the time by the bracing possibility of error and risk that are part of the experience and expectation of all machines. The fact that tuning to frequencies also involved numbers, the imaginary embodiment of the idea of that which is purely objective and untainted by human feeling or fantasy, was precisely what assisted their suffusion with desire, belief, and the desire of belief. The pseudo-complexity which opened up a kind of mathematical sublime in the idea of an entire universe made up entirely of matter vibrating at different rates or simply different rates of vibration constituting matter as such, was domesticated by the absurdly simplistic numerological idea that every entity in the universe has its own identifying frequency or call-sign. Number represents the horror of the undead (Connor 2016: 53-71) – which was precisely what made the vitalist turning of number to its own account by converting quantity to quality such an imaginary triumph. Call this the logic of immunity through imaginary jeopardy, a logic that we have already encountered in the consideration of medical machines.

One can perhaps point to another kind of tuning. The construction of imaginary machines often depends upon a mimetic principle in their naming and description, which are intended to win credence by the conspicuously mechanical nature of their rhetoric. Language and especially literary language is rich in opportunities for the deployment of what are rightly called rhetorical 'devices'. Copulative constructions like 'thanato-auric' can be particularly effective. Ben Jonson was already strongly attuned to the comic possibilities of alchemical pseudo-technicality in *The Alchemist*, but the huge explosion of technical terms and compounds from the 1890s onwards, verbal apparatus, one might say, for the naming of actual -graphs, -phones, -meters and -scopes, gave many opportunities for electro-verbal assemblages. Connectives carried the force of connections. The strong association of technical and engineering advances with Germany during this period also assisted this libidinisation of verbal compounds like the word 'radio-active', which in itself suggested the coupling of exactitude with generative energy. This display of elaboration is one of the many confirmations of David Trotter's connection, in *Paranoid Modernism* (2001), of paranoia, and especially the kind that involves what is called the 'technical delusion', or persecution by means of thought-controlling machines, with questions of professional standing and authority.

The disintegrator weapon devised by Edison in *Edison's Invasion of Mars* depends upon tuning employed for the purposes of aggressive targeting:

> Having obtained an instrument of such power, it only remained to concentrate its energy upon a given object in order that the atoms composing that object should be set into violent undulation, sufficient to burst it asunder and to scatter its molecules broadcast. This the inventor effected by the simplest means in the world – simply a parabolic reflector by which the destructive waves could be sent like a beam of light, but invisible, in any direction and focused upon any desired point.
> (Serviss 1947: 14)

As in Marie Corelli's *The Life Everlasting*, the fantasy machine represents the impossible combination of the two kinds of logic we have already encountered, that of a radiation that is not subject to any kind of limit or constraint, and the dream of being able to aim it. The fantasy of the ray gun involves subjecting the logic of the field to the logic of the line. All the way through the narration emphasises the linear function of what is called variously 'the vibratory stream' (82), 'the vibratory force' (83), 'the vibratory emanation' (85), and 'the vibratory current' (85). This reverses the evolution of machines evoked by Michel Serres in his essay 'Turner Translates Carnot', in which a fundamentally geometrical understanding of a world reducible to lines of force, and emblematised in George Garrard's 1784 picture of the warehouse of Samuel Whitbread, all hawsers, slings, levers, pulley-blocks, and derricks, and in Lagrange's *Analytical Mechanics* (1788), gives way to the swirling vortices of mist, smoke and fire of Turner's painting:

> Matter is no longer left in the prison of diagram. Fire dissolves it, makes it vibrate, tremble, oscillate, makes it explode into clouds. From Garrard to Turner, or from the fibrous network to the hazardous cloud. No one can draw the edge of a cloud, the borderline of the aleatory where particles waver and melt, at least to our eyes. There a new time is being fired in the oven. On these totally new edges, which geometry and the art of drawing have abandoned, a new world will soon discover dissolution, atomic and molecular dissemination. (Serres 1982: 58)

This means that the very possibility of representing such relations is put under threat:

> Garrard paints an exhibition, a dense tabulation, plane by plane, from the foreground to the background. Wright exhibits as well. The forge is still a theatre, and the painting could have served as an advertising sign. A work scene, the workers seen from behind, nothing is left to chance [...] There is no longer any representation in Turner's foundry. The painting is a furnace, the very furnace itself. It is a disordered black mass centered on the lighted hearths. We pass from geometry to matter or from representation to work. By going back to the sources of matter, the painter has broken the stranglehold of copying in the arts. No more discourses, no more scenes, no more sculptures with clean, cold edges: the object directly. Without theoretical detours. Yes, we enter into incandescence. At random. (61-2)

To reduce the force of radiation to a straight line which can be aimed like a firearm and selectively pick out a particular target to be annihilated is a counterfactual reduction of the field of force to a form. And yet, of course, that is precisely what was already in 1898 being done with X-rays after it became clear that radiation would destroy cancerous cells more rapidly than healthy cells, and would soon be done with radium and other kinds of radiotherapy.

One of the things that tends to distinguish literary from more obviously psychotic forms of technical delusion is that the latter are so entangled with the machinery of the body. David Trotter has pointed to the correlation between the influencing machine delusion and bodily nausea: 'It seems odd', he justly remarks of the persecutory mechanism of Natalija A., the patient on whose influencing machine Victor Tausk reports, 'that a persecutory system equipped with all the most up-to-date technology, as this one was, should have its way by the manufacture of snot and bad smells' (Trotter 2001: 68). Influencing machines, we might say, seem deeply preoccupied with 'effluvia', a word that connects the electro-etherial and the cloacal. And indeed, the experience of the body as an active machinery, especially in relation to its visible intakes and outputs, and concomitantly of machines as a kind of body, is very characteristic of those suffering technical delusion. The association of nausea with machine influence is suggested by the commentary that L. Heintzen produced to accompany the picture of the allegorical apparatus constituting his illness, which defined him as

> the pfuff object [*Pfuffobjekt*] of a nature-experimental philosophical pfuff society. A beast-person with an animal organ that serves the demon as a whore in order to breed vomited bestiality in the form of vomit, so that mania-vulva and dementia-vulva make me want to vomit in order to make a vomiter out of me out of the vomit. (Jádi 2006: 210)

Trotter's suggestion, that nausea is the paranoiac's way of defending against the machine's violation of the boundary between inside and outside, seems to propose that there might be two kinds of machine: a messy, in-between-states, glutinous corporeal machine that invades body boundaries, and a hygienic, cerebral machine of system that strives to restore and maintain them.

The psychotechnographic life of X-rays seems to embody this complexity. X-rays were seized on by psychics and supernaturalists as the ocular proof of various kinds of spiritual energy, odic force, or thought-stuff. And yet they were also from the beginning, in the famous 1895 photograph radiated by newspapers almost instantly round the world, if not at the speed of light, then certainly at the speed of new publishing media, of Bertha Roentgen's skeletal hand bangled with its tumorous wedding ring, associated with the inside-out horror of the excoriated or eviscerated body.

Yet, with X-rays, the flesh was delivered up, not as meat, but as light. The body revealed by X-rays was not merely irradiated, it was made radiant. Alongside those who saw the worldly applications of X-ray photography, in medicine, metallurgy and archaeology, there were many who saw in its results the proof of the spiritual or astral body. The *Herald and Presbyter* newspaper proclaimed that

> This discovery corroborates, so far as any material experiment can, Paul's doctrine of the spiritual body now existing in man. It proves, as far as any experiment can prove, that a truer body, a body of which the phenomenal body is but the clothing, may now reside within us, and which awaits the moment of its unclothing, which we call death, to set it free. (quoted Glasser 1933: 206)

Hippolyte Baraduc, a doctor at the Salpêtrière hospital who had been experimenting with the photographing of invisible soul-emanations for some years before the news broke about X-rays, saw in them not the shadows formed by the partial occultation of the rays, but rather a positive imaging of the soul – for 'light is [...] shadow is not' (Baraduc 1913: 74). X-rays went public just as Baraduc's book *L'Âme humaine* (1896) was in press, but he was able, writing of himself in the third person as was his practice, to add the following explanation of them:

> The interesting fact of procuring photographs of the hand showing its skeleton and its form, struck the scientific world with astonishment, it is the faculty which this invisible light had of lighting up the cavity of the body by illumining or by exciting, so to speak, the intimate and interior light of the fluidic body, which Dr Baraduc had iconographed two years before. The vital soul appears so luminous that, alone, the most opaque bodies which possess less luminous life, such as the bones, decide upon the spectral shadow of the totality of the organs: neither veins nor nerves appear, all is immerged in the intensity of the photochemical light of the animistic body. (77)

Baraduc saw X-rays as the way in which science had at last become 'acquainted with the luminous body' and therefore as 'a link between the purely physical known experiments and those of a more elevated order' (77-8). The semi-translucent mist of X-ray flesh resembled the bodily forms displayed by spirit photographs. X-ray flesh was therefore a kind of teleplasm. It was not the inert, dark body of the en-soi, but the soft body of the lived flesh, the flesh I inhabit and seem to myself to know as my own. The X-ray appeared to show the flesh ensouled, suffused with spirit. This was the aspect of X-ray vision that was taken up most enthusiastically by occultists and supernaturalists. X-ray vision revealed what occultists had been saying all along, that matter was fundamentally insubstantial, especially when regarded by the transpiercing inner eye. Swâmi Abhedânanda's *How to Be A Yogi* (1902) attributed the power of X-ray vision to the enlightened:

> They digest their food consciously, as it were. They claim that by a third eye they can, so to speak, see what is going on in their internal organs. Why should this seem incredible to us when the discovery of the Roentgen rays has proved everything to be transparent? (Abhedânanda 1902: 47)

This dream manifested itself in H.G. Wells's *The Invisible Man*, which was published in 1897, the year following the worldwide radiation of the idea of X-rays. Wells's hero employs the idea of being able to vary the refractive index of the materials of the human body, reversing the process in which glass when it is smashed and powdered loses its transparency:

> 'Just think of all the things that are transparent and seem not to be so. Paper, for instance, is made up of transparent fibres, and it is white and opaque only for the same reason that a powder of glass is white and opaque. Oil white paper, fill up the interstices between the particles with oil so that there is no longer refraction or reflection except at the surfaces, and it becomes as transparent as glass. And not only paper, but cotton fibre, linen fibre, wool fibre, woody fibre, and bone, Kemp, flesh, Kemp, hair, Kemp, nails and nerves, Kemp, in fact the whole fabric of a man except the red of his blood and the black pigment of hair, are all made up of transparent, colourless tissue. So little suffices to make us visible one to the other. For the most part the fibres of a living creature are no more opaque than water'. (Wells 1995: 83)

The process is achieved by a form of radiation, which is both coyly distinguished from X-rays and in the process associated with them:

> 'the essential phase was to place the transparent object whose refractive index was to be lowered between two radiating centres of a sort of ethereal vibration, of which I will tell you more fully later. No, not these Röntgen vibrations – I don't know that these others of mine have been described. Yet they are obvious enough.' (86)

That Wells should have found himself so taken up with uncomfortable reflections on exactly what would and would not be made visible of and in the body of the Invisible Man is one of the reasons why the story is aptly subtitled 'A Grotesque Romance'.

X-rays suggested more than the possibility of seeing into the secluded interior of the body. For many occultists, X-rays were an analogy – visible in their effects if not in their actuality – for the process whereby the body itself was believed to propagate force and form beyond itself. A parallel was commonly drawn in the 1890s between X-rays and the force of 'od' that was the subject of enthusiastic investigation by Carl von Reichenbach in the 1850s and 1860s. Reichenbach saw the od as a force radiated by all living beings, especially, of course, higher beings like, well, men:

> an uncommon degree of radiation is attributable to the force we call od, whose bounds, perhaps, like those of light, lie in the infinite. The consequence of this radiant energy is that we carry about with us continually an illimitable train of radiant light which, undetected by our own eyes, sweeps into space from our fingers, toes and limbs, and that, as living beings formed of matter, we are surrounded by a luminous atmosphere of our own, which we take with us wherever we go. (Reichenbach 1926: 91)

For these experimenters, the body was not merely the penetrated object of radiation, but also itself a radiating source. As Akira Lippit has suggested, '[t]he surface of the X-ray opens onto an impossible topography, a space that cannot be occupied by the subject or object. Or rather, a space in which the subject and object are dissolved into a phantasmatic hybrid or emulsion. (Lippit 2005: 80)

The problem for the paranoiac seems to be that the two machines, the hygienically hands-off machine of psycho-radionic fantasy, and the dirty, dabbling machine of visceral disordering, keep on collapsing into each other. The history of machines of radiation may be related as a gradual modulation of hard machines which are themselves palpable to the eye into soft apparatuses that may allow for heightened visualisation, but are themselves progressively withdrawn from visibility (Connor 2012: 212). But there is another sense in which radiation machines are allied to the Serresian soft, in that they bring to awareness the sense of the formless interior of the body; in the writings of a signature schizophrenic like Antonin Artaud, there is an alternation of the abstract rapture of the body-without-organs and the churningly abject mush of the body-that-is-all-organs.

The same interplay as Trotter makes out in paranoid influencing machines between the technical and the emetic, the analytic and the anal, arises in fantasies of alien invasion. Like the operators and operations of influencing machines, extraterrestrials are both technically hugely-advanced and physically retrogressive. The bodies of invading extraterrestrials are not so much terrifying, like sharks or tigers, as disgusting, like insects or sea creatures. Where they may seem to be all mechanism, like Daleks, they turn out to have nauseatingly squishy and slimy insides. Green, the normative alien colour, as well as the colour arbitrarily assigned to X-rays and early computer screens, is the colour of putrefaction. Alien abductors are usually possessed of advanced telekinetic, levitation, and dematerialisation kit, yet apparently still rely upon, or are lewdly addicted to, the use of anal probes.

Radiation often presented itself as a clean and abstract force. Serviss's *Edison's Conquest of Mars* indulges crudely sadistic appetites in its accounts of the zapping of the Martians, but protects itself from those appetites by the fact that Edison's disintegrator is a surgical weapon which has the capacity to cauterise the horrifying wounds it inflicts:

> Some of the giants had been only partially destroyed, the vibratory current having grazed them, in such a manner that the shattering undulations had not acted upon the entire body.
> One thing that lends a peculiar horror to a terrestrial battlefield was absent; there was no bloodshed. The vibratory energy, not only completely destroyed whatever it fell upon but it seared the veins and arteries of the dismembered bodies so that there was no sanguinary exhibition connected with its murderous work. (Serviss 1947: 85)

And yet the logic of the influencing machine is that it rends the wholeness of the body, so is on that basis an unclean matter out of place, or displacing matter. And the duality of ray and radiation machines is matched by the duality of the nausea that defends against incursion. For the problem with nausea is that it is so, well, nauseating, its convulsive, mechanically-produced emissions suggesting, not a body restored to balance and continence, but an exorbitant body unable to contain itself. Ever since the hoax of Jacques de Vaucanson's Canard Digérateur or Defecating Duck (Riskin 2003), the production of waste has been at once the sign of the living and the function that shows organic forms at their most ridiculously or disgustingly mechanical. The radioactive machine offers no simple defence against bodily abjection, since the non-object of radiation is itself so productive of abject ambivalence. It seems appropriate that the effects of radiation, which it had been easy to imagine in the beginning as a kind of universal soap, product of a kind of electrical Port Sunlight, should have become known as 'radiation sickness'.

Tausk's influencing machine has been taken as the anticipation of an emerging relation to a world in which machines are at once ubiquitous and immaterial. As such, it is a distorted reflection of the modern experience of mechanical transmission and reproduction – as technology gone bad, or mad. But perhaps the idea of the influencing machine is in fact a kind of theory of the relation between machines, minds, and bodies – not just a subject for psychotechnography, but itself a psychotechnography in the making. In 1919, L. Heintzen, a psychiatric patient in a hospital in Düren, produced a schematic diagram of his illness, which he entitled 'Allegorie meiner Krankheit'. Among the components of the diagram are three magnets, 'the frightful magnets of hell' (Jádi 2006: 2011), designated 'Despair, Restlessness, Delusional Ideas', 'Crime and Vice of All Kinds', and 'Suicide and Adverse Mental Disturbances'. There are also three regions marked as 'The dark rock of my past', 'The persecution' and 'The path through my illness to recovery, or through delusion to the light (unaware to me)'. Speaking tubes are also in evidence, described as 'The Mouthpieces of the Hereafter' (206-7). David Trotter is among those who have pointed to Freud's unease at the analogies between Daniel Paul Schreber's account of the delusive machinery of persecution and his own analysis of that account (Trotter 2001: 57). The form and idea of the influencing machine has itself become influential – contagious, proliferative, radioactive. The influencing machine is soft not just in the sense that it is immaterial, nor just in the sense that it softens what it encounters and penetrates, but also in the sense that it itself has no obvious outline or limit. The influencing machine tends to grow outwards, to assimilate or insinuate itself into any account that might be given of it.

Technical delusion is a disorder of reflexivity, in which it is the function of the influencing apparatus to mediate the self to itself. Appropriately enough, perhaps, given the accusatory content of the messages the influencing machine commonly transmits, it can be thought of as a traffic between the nominative and the accusative case of the subject. It is the means whereby the subject, so vulnerable to terrifying experiences of scattering or dissolution in other forms of psychosis, can return itself to itself as object. Technical delusions can sometimes seem like pathoplastic modulations of material that might in other periods been processed in demonological terms, as the pranks and oppressions of imps and devils, but one might say that the logic of technical delusions is parabolic rather than diabolic, since they seem designed to ensure that everything that goes out must bend back to its origin, like a radio signal that orbits the earth to arrive back at its starting point.

It is commonly noted that psychotic experience converges with contemporary experiences of '[t]echnology's dissolution of those barriers that protect the intimate sphere of a subject' (Fuchs 2006: 31). But parabolic logic is more complex, because more reflexive. This parabolic logic can help to account for the quality of incontinent excess that characterises paranoid self-documentation, and indeed the need for documentation at all. Images of influencing machines by patients such as Jakob Mohr and Robert Gie typically instance a *horror vacui* in their obsessive striving to fill every portion of the available space. Written accounts of technical delusions are similarly copious and unable to tolerate summary or abbreviation. Since her discharge from psychiatric hospital in 1995, Vanda Vieira-Schmidt has produced almost 50,000 pages of drawings responding to the plague she believed she had discovered of wicked people allied with the devil operating portable uranium devices, who assail their victims on the underground railway (Röske and Brand-Claussen 2007: 21-3). The radiative imaginary provides the driving force behind this diffusive energy of propagation, which seems to hunger to expand beyond every confinement or localisation. Its hunger is so great that, like the expanding universe, it seems to be driven by the desire to outrun its own capacities. This may help make sense of the rather odd remarks that Victor Tausk offers about the incompleteness that is typical of the influencing machine:

> The schizophrenic influencing machine is a machine of mystical nature. The patients are able to give only vague hints of its construction. It consists of boxes, cranks, levers, wheels, buttons, wires, batteries, and the like. Patients endeavor to discover the construction of the apparatus by means of their technical knowledge, and it appears that with the progressive popularization of the sciences, all the forces known to technology are utilized to explain the functioning of the apparatus. All the discoveries of mankind, however, are regarded as inadequate to explain the marvelous powers of this machine, by which the patients feel themselves persecuted. (Tausk 1933:520)

In an open universe, the machine can only be complete if it is, like the universe, incomplete, even as this hyperbolic expansiveness must never be allowed to escape the gravitational pull of the paranoid logic that can leave nothing unaccounted for. Hence the necessity for the mechanisms so painstakingly documented in Daniel Paul Schreber's delusions also to be self-archiving, by means of the *Aufschreibesystem*, or inscription system, and for Schreber's incapacity to recognise that it is in fact he who provides the means whereby the persecutory mechanism can keep going back on and going over itself. Since it is the role of the persecutory mechanism to keep him at a distance from himself, to prevent him collapsing into a black hole of nescient nonself-coincidence, it makes sense for it consume and command as much space as possible. This is perhaps why the radiative imaginary so often involves fantasies of space travel, for outer or interplanetary space can thereby figure space itself, the space that is both opened and filled by the imaginary machine that conducts the subject parabolically beyond and back to itself.

There is a widespread assumption that paranoid influence machines are to be understood as a kind of allegory or early-warning system regarding the technical or mediatic conditions of modern life. Dusan Hirjak and Thomas Fuchs conclude that

> schizophrenic self-experience displays a particular affinity to modern technologies and technical processes. Today, people's personal life and daily experiences are being highly influenced by new technologies as well. The concealed effects of the internet, television, radio, computer, electricity, cellular phones, and other contemporary technical innovations have generated alterations of our experience that to a certain degree resemble the schizophrenic way of being in the world: a dissolution of spatial and temporal boundaries, a virtualization of the world

and, last but not least, a reification of the human psyche. (Hirjak and Fuchs 2010: 101)

Alfred Kraus also sees the technical delusion as a kind of diagnosis of the new technical conditions in which '[s]pace, that is, being in different places, does not separate human beings from each other anymore [...] we can be continually present everywhere and reachable for everybody' (Kraus 1994: 59). This way of thinking misses the most salient and seemingly unignorable thing about the modes of technical delusion, namely that the machines, apparatuses, and processes they involve are *imaginary*. They are not documentations of any kind of simply existing experience, perhaps partly because they are themselves ways of having or bringing into existence experience that could not be known or shown in any other way; they are in fact productions rather than expressions. Jeffrey Sconce suggests that where '[p]aranoid schizophrenics have struggled for decades with the delusion that that they were suffering as coerced *targets* of invisible transmission', psychotics in the future 'will consider themselves to be willing *sources* of media content and transmission' (Sconce 2011: 89). But the elaborate machineries produced by subjects of influence like Corelli and Schreber are always in fact sources as well as targets. The extraordinary elaboration of technical delusions is a clear indication that they are the result of a form of imaginative labour, and that the space of the psychotic's representation is a kind of laboratory. If influencing machines seem to draw strongly on aspects of what I have been calling the radiative imaginary, their intention and effect is in fact to draw this diffusive machinery together into a single coherent machine, which, precisely to the degree that it constitutes a coherent system, must be imaginary.

This means that the imaginary machine must in fact be embodied in some secondary, graphic machine, consisting either of image or text, or often some combination of the two. There is agreement that sufferers from the influencing machine are particularly drawn to self-representation through writing, to the point where their self-reports can be regarded as an embodiment of their disorder. Burkhart Brückner analyses the voluminous 1108-page testimony to 'magnetic poisoning' of Friedrich Krauß as an instance of scriptive self-fashioning – Krauß being in fact a calligrapher as well as a writer, who described his calligraphy as 'the only talent not ruined by the magnetisers' (Brückner 2016: 30). Freud begins his essay on Schreber with the odd, indeed scarcely intelligible, defence of the use of textual analysis, rather than the methods normally prescribed in psychoanalysis, that

> [s]ince paranoics cannot be compelled to overcome their internal resistances, and since in any case they only say what they choose to say, it follows that this is precisely a disorder in which a written report or a printed case history can take the place of personal acquaintance with the patient. (Freud 1953-74: 12.8)

It is the drive to give the influencing apparatus objective existence as writing that furnishes the proof that this is an imaginary machine, precisely because it is a machine that is in fact identical with its figuration. The influencing machine believed to exist in the world is really just an allegory of the machine of writing that sums and summons it up. The writing machine figures an influencing machine that itself a figure for the machine of writing, each a machine designed to exert influence over the other. The allegorical machine is indistinguishable from the machine of allegory itself. That is, it is the reality of the machine that makes it imaginary.

It is in the influencing machine that imagination and machinery come most closely together. For the purpose of the influencing machine is to control thoughts, often in the form of images, which will always include images of its own workings. Indeed, one might have to say that the whole content of the influencing machine's transmissions is ultimately itself and its work of transmission. The influencing machine is a not-machine, or a super-machine; it is a machine the workings of which may be imagined, and it can be subjected to the work of imagination. Yet it is an imaginary machine too, and what is imaginary about it is precisely that it is more than a machine, that it surpasses the limits that are a feature of every machine. But, again, is this not in fact a feature of every actual machine, that it tends towards a kind of imaginary condition which would surpass every limitation? Perhaps it is the inbuilt tendency or aspiration of every imaginary machine to become identical with a machine of thought.

7
Invisibility Machines

Perhaps, of all imaginary technologies, invisibility machines are the ones that have been dreamt of for longest and yet seem furthest away from fulfilment. One of the earliest invisibility stories is told by Glaucon in Plato's *Republic*, in order to make the point that it is only the fact that we are visible to others and therefore subject to the constraints of law that makes us act justly rather than unjustly. In the story Glaucon tells, at *Republic* 2.359d-360b, a shepherd Gyges finds an opening in the earth, which contains a hollow bronze horse with doors, and inside it the body of a huge man who is wearing a gold ring on his finger. Gyges takes the ring and suddenly discovers he has the power to make himself invisible with it. He arranges to be sent to court to report on the state of the flocks and uses his power to seduce the queen, kill the king, and become ruler in his stead.

Stories of invisibility often involve forms of apparatus, in the form of magical objects, and mechanical methods for employing them. A number of the elements in the story of Gyges are very familiar, as Kirby Flower Smith points out: rings are often credited with magical powers, and bronze is often associated with the magical powers of artifice (Smith 268: 272). One feature of the ring seems especially notable, though, namely that it has to be operated. Invisibility is conferred by a reversible action, which can be thought of as an on-off procedure, the turning of the bezel, the collar in which a stone might be set (though no stone is mentioned), first inwards and then outwards again. Glaucon takes care to show Gyges experimenting with the ring in order to discover its mechanical principles, making his own narrative mimic the toggling operation:

> He was sitting down among the others, and happened to twist the ring's bezel in the direction of his body, towards the inner part of his hand. When he did this, he became invisible to his neighbours, and to his astonishment they talked about him as if he'd left. While he was fiddling about with the ring again, he turned the bezel outwards, and became visible. He thought about this and experimented to see if it was the ring which had this power; in this way he eventually found that turning the bezel

inwards made him invisible and turning it outwards made him visible. (Plato 1998: 47)

Cloaks and caps feature commonly in stories of invisibility not just because they do in actual fact disguise at least some part of the body, but because they have a binary operation, being able to be donned and doffed. In its reversible structure, this kind of invisibility magic resembles the fort-da game described by Freud, which he interprets as the desire of a young child to master absence, and in fact involves playing with invisibility (Freud 1953-74: 18.14-16). Attached as it is by a thread, the cotton-reel is not just alternatively *fort*, gone, and *da*, there, but is in fact still there when it is apparently gone, because under control, as though it were present but invisible.

The mechanical nature of this operation is suggested, not just by its formulaic nature and its repeatability, but also by the fact that it can be set in train by accident. That this is true of many magical procedures is one of the many parallels between magic and mechanics. The important thing about performing a spell is that it does not matter whether or not you mean to, or mean it when you do; like a machine, it works anyway, by itself. Magical thinking means being able or allowed to mistake wishes for reality, though this is perhaps not fully described by Freud's formula 'omnipotence of thoughts' (Freud 1953-74: 13.84-5), since magical thinking always requires the thinker to subject himself or herself to exceptionless laws of functioning. Indeed, in Freud's account, patients are sometimes unnerved by the consequences of their own omnipotence of thoughts, and by a capacity to control the world that is theirs while not being fully in their control. What is more, the controls and operation of the invisibility ring are themselves invisible, or at least easily disguised as the everyday action of fidgeting with a ring.

There are many different kinds of machine that have been employed to overcome invisibility. The microscope and telescope allowed objects to be seen that were invisible because of their reduced size or great distance. Intriguingly, among the things that that Robert Hooke and the Royal Society hoped to bring to view with new instruments and apparatuses of investigation like the microscope were the very miniature machineries of which they assumed nature was composed:

> *It has been their principal indeavour to* enlarge and strengthen *the* Senses *by* Medicine *and by such* outward Instruments as are proper for their particular works. By this means they find some reason to suspect, that those effects of Bodies, which have been commonly attributed to Qualities, *and those confess'd to be* occult, *are perform'd by the small* Machines of Nature, which are not to be discern'd without these helps, seeming the meer products of Motion, Figure, *and* Magnitude; *and that the* Natural Textures, *which some call the* Plastick faculty, *may be made in* Looms, *which a*

greater perfection of Opticks may make discernable by these Glasses; so as now they are no more puzzled about them, then the vulgar are to conceive, how Tapestry or flowred Stuffs are woven. (Hooke 1665: sig. g1r)

Impressed though he was by the 'strange kind of acting in several Animals, which seem to savour so much of reason', Hooke was persuaded by his observations to the view that 'those are but actings according to their structures', for, he reassured his readers, 'there are circumstances sufficient, upon the supposals of the excellent contrivance of their machine, to excite and force them to act after such or such a manner' (Hooke 1665: 190). Thus an 'outward instrument' of visualisation like the microscope finds itself redoubled in bringing to visibility the inward and invisible machinery of things (Ball 2014: 209).

Machineries of seeing multiplied their forms and settings. Anatomy theatres were built to function as instruments of visualisation, their funnelling structure acting as a kind of architectural lens to focus attention on the opening of the body's usually invisible interior. Cinematography made things visible that were invisible because of speed rather than scale, slow motion allowing one, with what Walter Benjamin called the 'dynamite of the tenth of a second' (Benjamin 2007: 236), to open up to view what was buried inaccessibly within ordinary duration, just as the surgeon's scalpel opens up the interior of the patient's body (233). Spectroscopy and radio telescopy allowed for things to be visualised that cannot be detected by human vision or do not emit visible radiation, opening the way for the multiplication of devices for visualising objects and processes that do not even belong to the order of the visible. Hooke and others may have been disappointed not to find familiar kinds of machinery in operation in the world below the threshold of visibility, but contemporary nanotechnology opens up the prospect of manipulating matter at these scales in order to engineer it, or even creating machines designed to operate invisibly, often in the human body, in a fulfilment of the idea of injecting a tiny submarine into a patient's body in the 1966 film *Fantastic Voyage*.

Gillian Beer has suggested that the realm of the invisible underwent a decided change during the nineteenth century. Despite, or even in part because of the huge advances in optical technology that took place during this period

> [t]he *invisible* thus became a site of debate and perturbation for later nineteenth-century people [...] The invisible, instead of being placidly held just beyond the scope of sight, was newly understood as an energetic system out of which fitfully emerges what is visible [...] Increasingly, the invisible came to declare itself as a condition within which we move, and *of* which we are, lateral, extensive, out of human control; worse, not amenable to analysis yet replete with phenomena. The invisible might

> prove to be a controlling medium, not a place to be explored;
> a condition of our existence, not a new country to be colonized.
> (Beer 1996: 85, 88)

The apprehension of this unseen environment was increasingly made possible through apparatus and instruments of ever greater sensitivity and precision, whether auditory or electromagnetic. The invisible announced itself through the effects and traces of these mechanisms rather than through visible indices.

The question of control points us to an important feature of machines of heightened visibility, namely that they have the power, not just to make one's immediate circumstances invisible – we need no reminding of this in an era in which pedestrians and drivers are at permanent risk of losing all awareness of their surroundings in their absorption in their devices – but also to create the illusion of one's own invisibility. The effect of many magnifying instruments is to make the viewer seem invisible, at least to themselves. They are like the voyeur at the keyhole, as evoked by Sartre:

> My consciousness sticks to my acts, it is my acts; and my acts are commanded only by the ends to be attained and by the instruments to be employed. My attitude, for example, has no "outside"; it is a pure process of relating the instrument (the keyhole) to the end to be attained (the spectacle to be seen), a pure mode of losing myself in the world, of causing myself to be drunk in by things as ink is by a blotter. (Sartre 1984: 259)

Voyeurs, their seeing wholly identified with their visual instrument, may seem to be swallowed up in the joyous field of their own vision, since everything they see is proof of their own absconding or abstention from visibility; this is why the sound of a footstep in the corridor floods them with the shameful awareness of their own body, suddenly stripped of its enraptured invisibility, as 'a pure *monition*, as the pure occasion of realizing my *being-looked-at*' (277).

The unsettling of visual authority by the apprehension of the invisible was accompanied by an invasion of the invisible within the field of the visible itself. It is this kind of invisibility mechanism which is the focus of this chapter: not the technological augmentation or production of visibility, but rather its subtraction by mechanical means. Such an ambition distinguishes invisibility machines from most other kinds of machines, which are designed to enlarge or extend existing human powers. Invisibility certainly does confer power, but in a mixture of active and passive modes, in that it makes one *able to be unable* to be seen. If all machines may be seen as essentially weapons of survival or of an assault upon necessity, as Oswald Spengler alleged (Spengler 1932: 11), then the tactics and technics of invisibility are weapons of blinding. This is in fact true in a double sense, for selective invisibility both blinds your adversary and blinds them to the fact of their blindness. It is vital for your defence or the success of your attack that your adversary knows neither what, nor that, they cannot see, so that your absence from their visual field is itself absent. In the case of the Sartrean voyeur, or observer absorbed in their observation, the invisibility of the self to itself is a kind of self-blinding. This might be thought of as something like the double operation of the Freudian theory of repression, in which the subject is held, first to repress a wish or impulse, and then to repress that fact of the repression. So the production of invisibility is the production of a reduction.

H.G. Wells's *The Invisible Man* (1897) provides some confirmation of this complex interlocking of aggression and passivity in the agon of invisibility. The Invisible Man himself, whose name we learn is Griffin, is characterised from the very beginning by an extraordinary propensity for irascibility rising to sadistic violence. It is easy to interpret his initial brusqueness towards Mrs Hall, the keeper of the Coach and Horses inn where he takes refuge, huddling inside the clothing and bandages he uses to cloak his invisibility, as a desperate fear of detection. But, as the narrative proceeds, Griffin becomes ever more unable to keep either his temper or the low profile that would be to his advantage. He seems impelled to be, as it were, exhibitionist in his invisibility. He snaps, swears, snarls, knocks over furniture; when the local doctor glimpses his flapping sleeve, he unaccountably insists on showing him the empty arm, to the doctor's consternation. The novel soon dissolves almost literally into a knockabout farce, with the villagers of Iping being kicked, tripped, pummelled and terrorised with pokers, as though by some devil or poltergeist. James Whale's 1933 film of the novel pushes this even further with Griffin signifying his presence by regular shrieks of maniacal laughter, accounted for by a bit of dialogue informing us that the chemical he has injected sends its subjects insane.

Griffin informs his confidant Dr Kemp that he has been actuated in his quest for invisibility by a kind of *folie de grandeur*, fuelled by resentment at his low and trivial surroundings:

> 'To do such a thing would be to transcend magic. And I beheld, unclouded by doubt, a magnificent vision of all that invisibility might mean to a man – the mystery, the power, the freedom. Drawbacks I saw none. You have only to think! And I, a shabby, poverty-struck, hemmed-in demonstrator, teaching fools in a provincial college, might suddenly become – this.' (Wells 1995: 84)

In a sense, the invisible Griffin is hemmed in by not being able to demonstrate his monstrous invisibility. And yet that invisibility also makes him vulnerable. The partial nature of his visibility, disclosed by accidental coatings, is often figured as a kind of scum or abjectly ambivalent state of matter (his nakedness means that for most of the novel he suffers from a cold, so even when he is not coated by drizzle or mud, his presence is suggested by phlegmy coughs and sniffles):

> I could not go abroad in snow – it would settle on me and expose me. Rain, too, would make me a watery outline, a glistening surface of a man – a bubble. And fog – I should be like a fainter bubble in a fog, a surface, a greasy glimmer of humanity. Moreover, as I went abroad – in the London air – I gathered dirt about my ankles, floating smuts and dust upon my skin. I did not know how long it would be before I should become visible from that cause also. But I saw clearly it could not be for long. (105)

Invisibility machines often work, not on objects in the world – as machines for lifting, transporting, sorting, multiplying, destroying, or converting energy do – but on another mechanism, the apparatus of human visual perception. One of the most important principles involved in the production of invisibility is that of camouflage, which confuses perception by encouraging the eye to see continuities rather than discontinuities. Seeing is cognitively an expensive operation, and, though alert to things that suddenly start out from a background, the eye has a predisposition for things that appear continuous and homogeneous, especially if they are, so to speak, complexly continuous, like the churnings of the sea or the dapplings of forest light. Camouflage techniques therefore encourage the eye to ignore the outlines of things.

The history of conjuring and what used to be called to be called 'juggling' anticipates the extraordinary discoveries of hypnotism during the nineteenth century, that, far from being simply automatic, visual perception is an ongoing construction, such that hypnotised subjects could be persuaded to abstract certain objects from the field of vision. In fact, there is no need for subjects to be put into hypnotic trance, as is demonstrated by the 'invisible gorilla' experiment performed by Christopher Chabris and Daniel Simons at Harvard in the late 1990s. Subjects were asked to watch a one-minute video of basketball players and carefully count all the passes. Halfway through the video, a female student dressed in a full gorilla suit walked on to the court, thumped her chest and walked off, occupying about 9 seconds of screen time; around half the subjects did not recall seeing the gorilla (Chabris and Simons 2011: 5-6). The understanding of the intensely directed and therefore highly selective nature of perception was materially assisted by technologies like photography and phonography which, by preserving everything, revealed the 'gappiness' of perception.

So an invisibility machine belongs at once to physics, physiology, and psychology. In this, it is itself, like many of the imaginary machines considered in this book, a psychotechnography, a device for rewriting the perceptual machineries of the self. An imaginary machine for making things invisible is a machine that works not only *in* but also in part *on* imagination, intervening in the process whereby we summon up and sustain our conviction what we see is the world, rather than our picture of it. It is this conviction of visual command that ensures that an image must always be imaginary.

Black Boxes

If machinery is implicated in the production of invisibility, the question of visibility is also to the fore in the experience of machinery, since there is in every machine a tension between appearance and operation.

It is necessary for every machine to have workings, moving parts that may be distinguished from each other, and for the ways to be indicable in which the moving parts work on each other. And yet, it seems to have become usual, sometimes for the more effective operation of the machine, but more often as a more obscure kind of necessity, for those parts to be modestly obscured from view. It is this which seems to make all machines magical. It is intensified by the tendency for machinery of all kinds to become smaller and less conspicuous, along with the development, in recent years, of distributed forms of technology that are much less likely to be physically apparent (Denning 2002). Ours has become 'an age of *technological invisibility*', writes Philip Ball:

> Thanks to the shrinking of electronic and mechanical technologies, the occult cogs, levers, and electrical switches of the machine – the same devices that once seemed to breathe life into ingenious automata in an indivisible blend of magic and mechanisms – have become truly too small to see, allowing them to be packaged into smooth-contoured bodies that do wondrous things without a visible means of action. (Ball 2014: 225)

It is perhaps because they act through workings that must remain concealed for them to continue to work that machines seem allotropes of the concealed automaton that is the human body. The ghost hidden in the machine is this quality of being hidden or withdrawn from view.

There is a certain obscenity about a machine that operates in full view, with everything that should be off-scene drawn into the field of the visible. It is the obscenity of the body opened up to view, or the obscenity of the writing machine in Kafka's 'In the Penal Colony', at the end of which the machine that has been described and displayed working so immaculately starts to vomit up its working parts:

> [H]e heard a noise from up in the Designer. He looked up. Was that cogwheel going to make trouble after all? But it was something quite different. Slowly the lid of the Designer rose up and then clicked wide open. The teeth of a cogwheel showed themselves and rose higher, soon the whole wheel was visible, it

> was as if some enormous force were squeezing the Designer so that there was no longer room for the wheel, the wheel moved up till it came to the very edge of the Designer, fell down, rolled along the sand a little on its rim, and then lay flat. But a second wheel was already rising after it, followed by many others, large and small and indistinguishably minute, the same thing happened to all of them, at every moment one imagined the Designer must now really be empty, but another complex of numerous wheels was already rising into sight, falling down, trundling along the sand, and lying flat. (Kafka 1993: 157)

The machine is in fact an apparatus for making manifest, for exposition. Its aim is to make the condemned man, his crime, and his sentence all identical, in a form that leaves nothing, as we say, to the imagination, by inscribing the judicial sentence as a corporeal sentence, in the body of the Condemned Man. But this making visible turns out to be dependent on the absence from view of the actual workings of the machine. The exposure of the machine somehow undoes all the work of the exposition, indeed, seems to make it palpable that there has been no machine at all, but only the vile, humiliating fact of torture:

> he bent over the Harrow and had a new and still more unpleasant surprise. The Harrow was not writing, it was only jabbing, and the Bed was not turning the body over but only bringing it up quivering against the needles. The explorer wanted to do something, if possible, to bring the whole machine to a standstill, for this was no exquisite torture such as the officer desired, this was plain murder. (157-8)

There seems to be a necessity for something we call a machine to be withdrawn from visibility. Often this will mean that the machine will work best if its smooth operation can be monitored better by the ear than by the eye – the thrum of the washing machine, the steady thud of the turbine, or the smooth whir of the propeller. Henry James was soothed by the click of the typewriter operated by his secretary Theodora Bosanquet as he dictated (Thurschwell 2001: 103), as though it were a guarantee of the regular operation of the machine of his imagination, an operation which would be subject to disruption if he were actually to try to imagine it. Sometimes, it is useful to the functioning of a machine that it should operate in a sealed or invisible environment; but for the most part there is no mechanical advantage to having the working parts of the computer, the television, the automobile, hidden from view. The invisibility of the machine seems to answer to some ulterior kind of necessity, some need for the machine to need to be imagined, or imaginary.

Machines, that is, operate on the principle of the black box, a contrivance where one can specify precisely an input and an output, without being able to describe precisely the process in the middle which results in the conversion. The black box principle is an important part of the magical thinking associated with all machines, whether they be factual or factitious. Alfred Abrams, who invented an imaginary form of energy-medicine called radionics, based on the careful adjustment of the balance between good and bad energy, instructed his customers under no account to open up his diagnostic and therapeutic devices, lest they interfere with their delicate workings – much as the Apple Corporation does today, perhaps for not entirely dissimilar reasons. When, in 1951, the US Food and Drug Administration opened up one of the $400 dollar machines marketed by Ruth Drown, one of Abrams's successors, they found only a simple electrical circuit, with no sign of any work to do for the nine dials displayed on the outside of the box (Young 1965: 159).

This ambivalence is a feature of Wells's *The Invisible Man*, which delays its revelation of the process whereby Griffin has become invisible until late in the narrative, the usual convention being that the exposition of transformative machinery occurs close to the beginning of such stories. Wells is coy, or perhaps just undecided, about what kind of mechanism has been involved in producing Griffin's invisibility. He tells us that, after leaving University College London, he 'dropped medicine and took up physics' (Wells 1995: 81). His experiments in changing the refractive index of organic matter seem to involve a lot of chemistry, with much compounding and talk of 'filtering'. Ingestion of certain substances seems to be enough to render most of Griffin's body translucent. And yet Wells also feels the need to invoke a bit of radioactive apparatus, which will work on the body from the outside rather than the inside, as in Griffin's explanation to Kemp:

> 'the essential phase was to place the transparent object whose refractive index was to be lowered between two radiating centres of a sort of ethereal vibration, of which I will tell you more fully later. No, not those Röntgen vibrations – I don't know that these others of mine have been described. Yet they are obvious enough. I needed two little dynamos, and these I worked with a cheap gas engine. My first experiment was with a bit of white wool fabric. It was the strangest thing in the world to see it in the flicker of the flashes soft and white, and then to watch it fade like a wreath of smoke and vanish.' (86)

As he embarks on his career of invisibility, Griffin is oddly determined to leave no trace of the means he has employed to achieve it, explaining to Kemp that

> 'it occurred to me that the radiators, if they fell into the hands of some acute well-educated person, would give me away too much,

and watching my opportunity, I came into the room and tilted one of the little dynamos off its fellow on which it was standing, and smashed both apparatus.' (92)

The invisibility machine once disposed of, it itself remains invisible, recoverable only through the three manuscript books in which Griffin has recorded his secret process in cypher – which are the last thing of which we gain a glimpse as readers.

Despite the fact that his invisible condition often yields Griffin's bodily interior up to view, ingested food remaining visible as a kind of floating slurry until he has fully absorbed it by digestion, and cigar smoke lighting up his bronchial tree like an X-ray, there is very little of what we might think of as Griffin's own interior experience or perspective. We remain entirely dependent on external evidence of his presence and intentions. This makes his invisible physiology entirely a matter of physics – of impacts and abrasions. The mixture of the comical and the queasy this produces may explain Wells's subtitle for the novel, *A Grotesque Romance*. Though magical machinery is largely kept offstage in the novel, almost all of the action may be thought of as mechanical, in an anti-Bergsonian sense that seems to make persons and objects indistinguishable. In this the novel surrenders the traditional advantage of interiorised narrative, subjecting itself to the constraint that we may know only of what may be shown, so that we never hear or see Griffin except through his effect on others. Indeed, not only does Griffin escape the gaze of characters in the novel, he even seems, in one extraordinary passage at the beginning of Chapter Two, after Griffin has fled Kemp's house, to give the narrative itself the slip:

> The Invisible Man seems to have rushed out of Kemp's house in a state of blind fury. A little child playing near Kemp's gateway was violently caught up and thrown aside, so that its ankle was broken, and thereafter for some hours the Invisible Man passed out of human perceptions. No one knows where he went nor what he did. But one can imagine him hurrying through the hot June forenoon, up the hill and on to the open downland behind Port Burdock, raging and despairing at his intolerable fate, and sheltering at last, heated and weary, amid the thickets of Hintondean, to piece together again his shattered schemes against his species. That seems the most probable refuge for him, for there it was he re-asserted himself in a grimly tragical manner about two in the afternoon. (119)

In subjecting his narrative to this arbitrary and unnecessary constraint, Wells seems to be anticipating the medium that will provide the most paradoxically recognisable and, in contemporary parlance, 'iconic' embodiment of the

Invisible Man, that of cinema. Indeed, in a certain sense, we may say Wells wrote his story of invisibility as a kind of screenplay for a kind of cinema that still lay in the future. This is not as surprising or clairvoyant as it might seem, for there had been a fascination with invisibility in the many machineries of image-making that anticipated cinema. Painting had developed many ways of using the visible to indicate the invisible, one of the most subtle and suggestive being Velasquez's ostentatious presentation of the viewers of his *Las Meninas* with the back of a canvas, on the concealed front of which we presume is the portrait of the royal couple in front of him who, we then realise with a ghostly start, must standing in precisely the place that we occupy to look at the painting. As we see the faces of the king and queen reflected in the mirror on the back wall, we surmise that they are a double of the faces being depicted on the concealed front of the painting, in which we also see the fact of our own invisibility to everyone in the painting.

It seems that it is only through the supplementary production of images that are both there and not there, in the form of paintings, theatrical performances, photographs, and films, that invisibility can be figured as a positive and visible absence. This may help explain the fact that technical means have often been employed to create effects of vanishing or power invisibly exercised. For the plot-resolving power of the *deus ex machina* to descend from the flies, rather than from stage left or stage right, signifies an incursion into the visible action from some other dimension, which is essentially rather than contingently invisible. In fact, it does not seem as though any effort was made to disguise the appearance of the *mechane*, or crane, which lifted or lowered gods, or important mortals like Socrates in Aristophanes's *The Clouds* (Brockett et. al. 2010: 10). The fact that one could see the mechanism perfectly well is itself a kind of instruction to the eye that it should be regarded as figuring the properly invisible; it is the lack of camouflage that is telling. The underneath of the stage was also a kind of reservoir of invisibility, from which demons and spirits could arise and into which they could depart, either through what the Greek theatre knew as the 'steps of Charon', or through trapdoors (Campbell 1923: 65). The projection of sounds and voices from beneath the stage in nineteenth-century ventriloquist performance recalls this tradition, confirming the convention that the space to the left and right of the stage is to be regarded as a simple extension of it, whereas the space above and below it is to be regarded as ontologically distinct, and a different dimension of existence (even where, as in the medieval theatre, there might be a hellmouth leading down to the underworld). The machinery of visualisation centres on the laborious operation of cranking weight through vertical distance. In his *Ten Books on Architecture*, Vitruvius even distinguishes between 'machines' (*machinae*), which 'need more workmen and greater power to make them take effect', and 'engines' (*organa*), which 'accomplish their purpose at the intelligent touch of a single workman' (quoted Campbell 1923: 18). Cloud machines that allowed Christ to ascend into glory were employed in medieval plays like the Towneley and N-Town Ascension plays (Davidson 1996: 97). Such machines were still commonly employed in the Renaissance. Isabella D'Este described a presentation of the Annunciation in Ferrara that impressed her by the fact that it showed God surrounded by a choir of angels in mid-air: 'No support could be seen either for His feet or for those of the angels, and six other seraphs hovered in the air, suspended by chains' (60). The principle of these machines is precisely that they remain invisible, suggesting the divine entering into the sphere of the mortal. Photographic media employing stop-action procedures were able to produce emergence from concealed places within the visual scene, rather than above or below it, as well as a kind of visible invisibility. Hard, or kinetic machinery therefore gives way to optical machinery, which in turn gives way to digital animation.

The phrase *deus ex machina*, translating Greek ἀπὸ μηχανῆς θεός, was commonly employed to refer to plays of Euripides, which use the device freely. By the time that Aristotle used the phrase in Book XV of his *Poetics* (Aristotle 1995: 81), it had moved from actuality to abstract device, signifying any sudden and unexpected resolution. The *deus ex machina* device has come to mean a slightly crude, or contrived summoning up of some hitherto hidden principle to untangle what might be an otherwise inextricable kind of confusion. The very elaboration of the phrase may seem to enact this disproportion between physical and mental effort. On the one hand, the resolution comes from nowhere, out of nothing, out of the blue. On the other hand, it is clear (or, so to speak, clearly concealed) that a great deal of work has been required behind the scenes to create the illusion. Invisibility in theatrical terms therefore seems to institute an economy between the something-for-nothing of a denouement that has simply and costlessly been dreamed up, and the expensive contraption required to give this impression.

The cinema took over many of the devices and machineries for creating theatrical illusion developed over the nineteenth century, which themselves increasingly employed optical as well as mechanical means. The best known of these illusions was 'Pepper's Ghost', first devised by the engineer Henry Dircks (who among other interests collected examples of perpetual motion machines, as we will see in Chapter Eight) and the Director of the Royal Polytechnic Institution, John Henry Pepper. An adaptation of this device was the 'Proteus' cabinet, invented by Thomas William Tobin, which used mirrors to allow somebody to appear to disappear from within a sealed cabinet. The principle here is not that something is added to the visible scene, but that it appears to be taken away, through a reflection that in fact covers the presence of something present (Steinmeyer 2005: 77) Tom Gunning, arguing for the roots of cinema in the possibility of visible disappearance, describes this as a kind of 'Copernican revolution' in optics, in that '[r]ather than conjuring a spirit, the trick made emptiness visible' (Gunning 2012: 59). The trick allowed for complex interlacings of appearance and disappearance, as this description from a performance at the London Polytechnic indicates:

Most interesting of all is something indifferently described as 'The Wonderful Cabinet,' and 'Proteus; or, we are here, but not here.' This is a new invention by Mr. Tobin and Professor Pepper. A handsome, cabinet is produced, large enough to contain two or three people upright; but is opened, and found to be empty. On the evening when we saw it, Mr. King desired Mr. Tobin to enter. He did so, and immediately after somebody else came from it in the costume of Paul Pry. Mr. Tobin had disappeared. Mr. Tobin again entered, and when the door was opened, there was a chattering skeleton with fragments of Mr. Tobin's attire upon him, thus leading to a certainty of the identity of the skeleton upon earth. Mr. Tobin reappearing at the side, was desired to enter the chest once more, to fetch his bones out, and then fresh complications occurred, as they did also with Mr. King, who certainly must have been in the flesh, as he was talking all the time. However, there is no 'must' at the Polytechnic; there is 'nothing certain in life' at the Polytechnic. ('Public Amusements' 1865: 8)

It is as though the power of mastering absence enacted in the fort-da procedure described in Freud's *Beyond the Pleasure Principle* (Freud 1953-74: 18.14-16) were autonomised and transferred to the visual scene itself – as though the visible itself could blink, or close its eye upon itself. But then, that is just the way in which the variant of the game reported by Freud works: rather than throwing a cotton-reel away and pulling it back into view, Freud's grandson would operate his own ring of Gyges, ducking down below the mirror and then coming back into sight.

The machine which best demonstrates the tension between display and concealment is the Mechanical Turk. This was a contrivance invented by Wolfgang von Kempelen in 1769, in which the mechanical figure of a Turk seated over a cabinet appeared to play games of chess. After having been displayed in a number of European cities, it was acquired in 1805 by a showman and engineer called Johann Nepomuk Maelzel. Having sold the automaton, Maelzel repurchased it in 1817 and took it on tour in the United States. Despite the fact that von Kempelen was a maker of genuine automata, having to his credit a machine for producing vocal sounds by artificial means, it was actually, as many suspected throughout its career, a hoax, in that it was not a machine at all. The Turk was operated by a succession of presumably well-remunerated chess masters hidden inside the cabinet.

Edgar Allan Poe said of the Turk that 'we find every where men of mechanical genius, of great general acuteness, and discriminative understanding, who make no scruple in pronouncing the Automaton a *pure machine*, unconnected with human agency in its movements, and consequently, beyond all comparison, the most astonishing of the inventions of mankind'

(Poe 1836: 318). The machine is pure because it is supposed to operate without human intervention. It is invention without intervention, without anything coming between the machine and its operation.

There is no machine. And yet there is a machinery for the production of the conviction that a machine is at work. The machinery of the spectacle, the making visible of the machine, is all intervention, in that it is insinuated between the seer and the seen. Poe carefully explains the procedure by which the machine is presented:

> Maelzel now informs the company that he will disclose to their view the mechanism of the machine. Taking from his pocket a bunch of keys he unlocks with one of them [...] and throws the cupboard fully open to the inspection of all present. Its whole interior is apparently filled with wheels, pinions, levers, and other machinery, crowded very closely together, so that the eye can penetrate but a little distance into the mass. Leaving this door open to its full extent, he goes now round to the back of the box, and raising the drapery of the figure, opens another door situated precisely in the rear of the one first opened. Holding a lighted candle at this door, and shifting the position of the whole machine repeatedly at the same time, a bright light is thrown entirely through the cupboard, which is now clearly seen to be full, completely full, of machinery. (320)

Maelzel follows conjuring convention in seeming to act out a kind of logical demonstration, in order to convey to the spectators the conviction that they have 'beheld and completely scrutinized, at one and the same time, every individual portion of the Automaton' (320). Everything is laid open to view, leaving the conclusion that the machine consists of nothing but its machinery. In fact, as Poe goes on to demonstrate, the device contains a chess player, who is trained to move his body into different postures as different doors are opened which allow the illusion to be given that the machine is machinery and nothing but machinery. This includes mirrors 'so placed to multiply to the vision some few pieces of machinery within the trunk so as to give it the appearance of being crowded with mechanism' (323). The 'direct inference' Poe makes from this is 'that the machine is not a pure machine. For if it were, the inventor, so far from wishing its mechanism to appear complex, and using deception for the purpose of giving it this appearance, would have been especially desirous of convincing those who witnessed his exhibition, of the *simplicity* of the means by which results so wonderful were brought about' (323).

In all of this, Poe himself relies upon what he calls 'demonstration'. That is, he is following, and mimicking, the logic of the original demonstration. 'It is quite certain', he writes, 'that the operations of the Automaton are regulated by *mind*, and by nothing else. Indeed this matter is susceptible of a mathematical demonstration, *a priori*' (319). But where Maelzl's demonstration, itself a kind of theatrical apparatus, is designed to produce the idea of a machine, Poe's demonstration is designed to show that not everything has in fact been laid open to view. Poe aims to demonstrate the machinery of dissimulation, the machinery by which the Mechanical Turk convinces us that it is a machine.

But this means that there is in fact a deeper analogy between Maelzel's operation and Poe's. In both cases, demonstration is used to show definitively that there is nothing of the workings of the machine that cannot be seen, and yet that there is also something that cannot be seen, namely how the machine works. For Maelzel, this is a machine that works in secret; for Poe, it is the secret that it is not the machine that works at all, but rather a mind. The Maelzel-machine and the Poe-machine are meshed together in an act of writing that must itself be a 'pure machine' of logical inference if it is to demonstrate that the machine is in fact pure mind – or rather, must convince its reader that it is such a machine, and not in fact the staging of its demonstration. In both cases, the machine is at once laid open to view, and operates in concealment.

There is an even more surprising form of black-box concealment, in which the operation of the machine is not hidden from view, but is rather exhibited, yet with a complexity that seems entirely incomprehensible. This illegibility allows the machine to be at once visible and withdrawn from view: one sees but does not know or cannot say with it is that one sees, meaning that once cannot be sure whether one is viewing hoax or authentic artefact. Some of the most mysterious machines found in art are of this kind, for example the strange apparatus devised by the inventor Canterel in Raymond Roussel's *Locus Solus*, which combines a machine for instant and painless extraction of teeth with an aerial device able to predict and exploit every change of wind direction, the whole operating to effect constant adjustments to a mosaic of discarded teeth of different colours (Roussel 1983: 24-36).

In 2005, Amazon borrowed the Turk's name for a system it made available to bring together employers and workers on Human Intelligence Tasks that human beings find relatively easy, but are difficult or expensive to automate, such as classifying images, editing and transcription of podcasts, and even searching image databases for traces of missing persons. The system was originally devised in order to enable individual workers to identify which of the millions of pages produced by Amazon were duplicates, a job which surprisingly was difficult to automate. In this kind of crowd-sourced microtasking, the workers are the 'Turks', or 'Turkers', who enable the machinery to work. They are, in the words of Jeff Bezos, 'artificial artificial intelligence' (Cushing 2013: 2). Turkers earn tiny amounts of money, which makes it odd that so large a proportion of them should be from the US (the other largest source of Turkers is India, which makes greater economic sense). In fact, Mechanical Turk is a symptom, a largely hidden one, in that it is not much known about or discussed, of a much larger and more structural dependence of artificial systems of intelligence on human ones. Many internet businesses are finding ways to make use not of labour time but of the data by-products of our other activities. We imagine a world in which more and more human actions and interactions are replaced by automated ones. A large part of the reality is that we are in fact working the machine, only, unlike the inmate of Maelzel's apparatus, we don't know how – or even that – we are doing it. This then becomes artificial artificial artificial intelligence, the black boxes that dissimulate the work forming multiple encapsulations.

Technologies of visualisation like phototography and cinema have displayed an elective affinity with machinery. The many films that deal with the making of life, or with statues, dolls, toys, robots, or machines that come to life, seem often to be a ceremonial enactment of cinema's own power to animate. The scene in which the monster is animated in James Whale's *Frankenstein* (1931) musters an unaccountable collection of art deco machines involving dials, phials, coils, and electrostatic tubes, up and down which sparks flash and ripple, along with, suspended over the bed on which the insensate monster is lifted up to the thunderstorm, a wonderfully purposeless barbell arrangement, suggestive of the Van de Graaff generator, the first model of which had in fact been demonstrated by Robert Van de Graaff in the very year the film appeared. The equipment in the laboratory is a confection of different styles and epochs of technology, from the winch that is laboriously employed to raise the monster on its bed to the roof, to the space-age globes. All of this in fact seems to embody not so much the power of machinery as the machinery of exhibition, which for centuries had been electricity's primary use (Morus 1998). The sparks that fly out intemperately suggest immense power under the control of human genius. The scene perhaps recalls the photographs of Nikola Tesla taken by Dickenson Alley for the *Century Magazine* in December 1899, in which double-exposure was used to show Tesla calmly reading amid a torrent of sparks streaming from an enormous transmitter (Carlson 2013: 297). And yet, of course, the real purpose of the sparks is to suggest a power beyond power that can scarcely be contained, not even perhaps in the space of vision, which is subject to blitzing white-outs as lightning flashes.

If there are suggestions in this scene of Lazarus being lifted on his bed to the roof, the raising of the monster to the heavens also suggests sacrifice and the forms of religious display known as 'ostension', a sort of exhibition to the second power, in a showing that shows itself. The monstrous may be that which is horrifying or sublime in its formlessness, but may also bring with it the recollection of the monstrance, sometimes also known as an ostensorium, a ceremonial holder in which the host or holy relics may be displayed to the faithful. The monster, from Latin *monstrare*, to show, is indeed a showing or revelation.

James Whale's transformation scene is affectionately parodied in Mel Brooks's *Young Frankenstein* (1974), which actually employed the props used in

Whale's film. We might say that cinema uses the scene of the laboratory in a similar way, to make visible the work of animation or transformation that is otherwise kept invisible in the work of visualisation.

So there are machines for making the invisible visible; and there are machines for making the visible invisible. There are also machines designed for seeing or visualising invisible mechanisms, along with machines that are themselves invisible. And there are machines whose visibility is itself a form of concealment in plain sight, or a way of making themselves, or part of themselves, invisible. The categories are not wholly distinct, and machines may move between these different conditions of, and relations, to visibility. Perhaps there might even be a kind of machinery in the logic that governs such motions. We may be sure at least that, through all kinds of machinery whatsoever, visibility is always at issue. A mechanism is always something capable of exposition, since to say something is mechanical is to say that its workings can be seen and shown. And yet to imagine a machine is always to imagine some occultation, working against, or even within the ocular, if only because our seeing or fantasy of seeing is always in part caught up in the workings of any machine we imagine or inspect. We can never, it seems, fully see what it is that we see in machines.

8

Perpetual Motion

As the titles of the previous chapters have indicated, machines are for doing things with. Whether it is for conveyance, measurement, communication, the giving of pleasure or waging of war, every machine is devoted to a purpose, often, though not invariably, one that may be achieved by some other means. Machines are in this sense extropic, turned away from themselves. To be reduced to the condition of a machine is thought of as a reduction, not so much because machines are dead objects as because machines are subordinate, surrogate, and instrumental, means to an end rather having any end in themselves, and so excluded from the Kantian condition that 'all rational beings stand under the *law* that each of them is to treat itself and all others *never merely as a means*, but always *at the same time as an end in itself*' (Kant 2012: 45). This is one of the simpler ways in which all machines can be thought of as media, in that they come between their users or originators and those users' purposes, as well as providing substitute ways to achieve those purposes, saving the users from having to do it for themselves. Machines are characterised in short by *dativity*, by the fact of being a *being for*. Things can become more complicated when machines start to seem capable of acting autonomously, but schemes to develop ways of behaving ethically toward machines such as those considered under the rubric of 'machine ethics' (Anderson and Anderson: 2011) do not on the whole modify this condition, but rather apply it differently.

This extropic orientation also underwrites modern machine aesthetics. Almost all machines allow for gains in efficiency with respect to the actions for which they substitute. The history of technology is the history of the search for ways to improve on improvement, and this in turn helps to form the principle that machines should not only be for doing things, but also, optimally, *only* for doing those things and nothing else. A maximally-efficient machine should be entirely consumed in its use, with nothing incidental to it or left over from it; thus a machine aesthetic is an aesthetic of pure function. The view has developed, in short (the shortness being consonant with the principle), that a machine should be only and entirely what it does.

Anomaly can, however, arise. On the one hand, no machine has its meaning or purpose in itself. On the other hand, a machine's quality of rigorous self-coincidence in fulfilling its purpose seems to give it a kind of self-identical integrity that other things, Kantian rational beings, or living organisms in particular, do not have, precisely because they can never be wholly subordinate to some other purpose. Machines can be essentially themselves precisely because of their subordination to some extrinsic purpose: they can be fully themselves because they are not their own masters. This tension between extropy and identity exerts its pressure on the word 'automaton', which originally meant moving by itself, but, under the influence of a history in which more and more automata and automatic systems of all kinds have been devised by humans to effect human purposes, has come to mean something like actuated and directed by some external power. So, to act like an automaton is to act without freedom, will-lessly, like a robot.

Following a suggestion made by Marvin Minsky, Claude Shannon built a machine that seems to exhibit, even to inhabit, this paradox. The machine consists of a box on which is mounted a single on-off switch. When it is switched on, the lid of the box opens, to let out a 'hand', or mechanical equivalent, which flips the switch to turn the machine off and immediately retreats into the box. It is often referred to nowadays as the Useless Machine, though Shannon's name for it was the Ultimate Machine. It is in fact the latter precisely because it is the former. Having no ulterior purpose, it scarcely at all meets the criterion just articulated for being a machine, that it be for something. If it is a machine for anything it is for turning itself off, so it is therefore a machine designed not to be a machine, or to thwart every effort to make it work as one. Yet it meets another criterion for being a machine in that its operation is indeed entirely taken up in its function, and its function entirely exhausted in its operation, without excess or remainder. Because its function is single-mindedly to fail to be what it is (a machine) and so also to be the machine that it is not, its being and doing seem indistinguishable. In fact, this is not entirely true, since most versions of the machine have comic or creepy adornments designed to assist the illusion of some slumbering entity inhabiting the apparatus which resents being woken into being. However, it would be perfectly possible for the machine to operate without these uncanny extras, since the machine is essentially nothing more than the material enactment of an algorithm.

This is perhaps one sense in which it is the Ultimate Machine, that it has removed every superfluity or inefficiency. But there is another, which connects with Shannon's preoccupation with cybernetic machines: machines able reflexively to monitor and modify their own conditions. Here is a machine that performs no other action other than to monitor and modify its own state. It is consumed by self-relation. It may be this which gives it a kind of uncanniness that is not possessed by a thermostat or a car equipped with cruise control. It seems to have the quality of being 'for-itself', to use the term that Sartre usually reserves for living creatures, in its anxious care for itself, or rather its care for its desire not to be.

The ethic of sleek self-coincidence has by no means always applied to machines. Jonathan Sawday writes of Renaissance engineers that '[q]uestions such as "will it work?" and "Will it perform to its design specification?" and even "Is it efficient?" were not the first, or even the last questions they asked of their designs' (Sawday 2007: 97). Renaissance machinery, which was sometimes, as in the inventions of Leonardo, the stuff of fantasy, and even when real had an element of the fantastical, was often designed to amaze and fascinate as well as to perform its function efficiently. Since the Renaissance, another kind of excessiveness has prevailed in which there is less than meets than eye rather than more. As we have seen, a certain kind of enigma seems to attach to machines, the workings of which can often seem unintelligible or occult. By no means all machines accord with the principle of complete and transparent self-evidence and indeed the Ultimate Machine itself only shows its function, not its workings, the wires, coils, springs, etc., which remain concealed inside the box.

This is the sense in which the Ultimate Machine might be regarded as in fact unable to achieve a condition of pure and absolute self-reference. For in fact it is not just a machine for switching itself off. At the very least it seems designed to produce a smile: beyond this, it may have a pedagogic, even an allegorical function, making it a machine for disclosing certain features about machines, and so intended to produce, not mechanical, but discursive effects (these current reflections, for example). This is another way in which it resembles many Renaissance machines: as Jessica Wolfe has observed, '[a]s intellectual recreations or conceptual tools for philosophers, scholars, and artists, Renaissance machines are only secondarily regarded as objects with specific functions and aims' (Wolfe 2004: 237). Whether they display or conceal their workings, there seems always to be a certain theatricality associated with machines (Knoespel 1992). This may be true even of machines that do not seem ostentatiously complex, for the simplest, most stripped-down machine may make or be taken to provide a meretriciously extravagant display of its own functionality. Such a machine may, to use Wittgenstein's terms, constitute both mention and use, since it will always be quoting itself, mentioning as well as performing its use.

There is one category of machine in which these complexities with regard to doing and showing, efficiency and exemplarity, are more prominent than in any other: the long dreamed-of, frequently-instanced *perpetuum mobile*, or perpetual motion machine.

The ontological status of perpetual motion machines is perplexing. In a sense, all such machines are imaginary, since it has seemed for some time that perpetual motion machines cannot exist. Indeed, perhaps it is the only kind of machine of which we can safely say that it will never exist. And yet perpetual motion machines are rarely merely what-if or in-principle projections, like teleportation devices, or anti-gravity fuel, or ray-guns, since the whole point of a perpetual motion machine is to show its workings, to be a working model of how a *perpetuum mobile* might in practice be able to work. In that they typically stand, fully apparent (or apparently so) before us, whether in a blueprint or working model, we may say that such machines are not so much fictional as fallacious. So perpetual motion machines tend towards a strange kind of ambivalence. They are imaginary, but not because they are unreal. The machines are real enough, and have often enough actually been constructed, but their alleged powers and effects are imaginary. In the case of Leonardo, who often seemed disinclined to believe in the possibility of perpetual motion machines, yet also devoted much time to investigating them, the point of this detailed designing may have been in part to understand the reasons for their impossibility, which required him to work through different proposals (Olshin 2009: 1-2).

Such machines work by showing they cannot work as they seem to or are supposed to. They follow what might be thought of as the logic of the hypochondriac – who is quite sure she is ill, and also quite correct in thinking so, but mistaken about the nature of her illness, suffering as she is, not from cancer or subcutaneous worms, but from hypochondria. The perpetual motion machine similarly always works, just never in the way in which it is supposed to. So, although perpetual motion machines seem entirely taken up in their own workings, existing only in order to perpetuate their own existence, their by-products, or accessory effects, may in fact often be the primary work they do. The work performed is often that of moral exemplification. Perpetual motion machines are didactic, allegorical, pedagogic, homiletic, a concentration of rhetorical rather than kinetic energies. Sometimes, their lesson can seem optimistic. Thus the emblem for 'Artifice' in Cesare Ripa's *Iconologia* (1593), which shows a man with one hand pointing to a hive of bees and the other pointing to a perpetual screw mechanism, demonstrates the concord between human ingenuity and the endless powers of the universe (Sawday 2007: 197). The perpetual motion machine could provide a warning against idolatry, but could also be presented as 'a godlike apparatus that would replicate the divine skill with which God had fashioned the universe' (118). Sawday sees the perpetual motion device exhibited by Cornelius van Drebbel at the court of James VI in the early 1600s as a defence of a pre-Copernican view of the universe revolving around a fixed earth (121). During the eighteenth century, perpetual motion machines, or plausible attempts at them, continued to be used to magnify the powers of princes and monarchs; but, as Simon Schaffer has shown, they also imaged the machinery of state, diplomacy, and the workings of credit in markets that seemed to some to be able to run on nothing. So, in the London stock market of 1720, '[a] mechanism which offered the prospect of extraordinary gain from an allegedly ingenious principle to which none were allowed access was a remarkably accurate picture of contemporary metropolitan reality' (Schaffer 1995: 182). There were rumours in 1721 of a project for a perpetual motion machine that had accrued investment of £1 million. The analysis of perpetual motion schemes was an important part of the process whereby secure ways of measuring work, investment, and value were developed. Perpetual motion machines were themselves emblems of the power of the human mind to accord credit to chimeras of its own making: 'It seems self-evident that neither nature nor technique can generate endless profit, so faith in perpetual motion is seen as a parable about the fallibilities of the human mind rather than about the capacities of technique' (159).

By the end of the nineteenth century, the *perpetuum mobile* had largely become an emblem of the *ignis fatuus* of scientific enthusiasm, the remnant of a magical relation to the idea of human artifice that belonged with alchemy and the construction of automata and had been swept away by the age of thermodynamics and the electric telegraph. The theoretical basis for this was the formulation of the Second Law of Thermodynamics, in the work of Sadi Carnot, Rudolf Clausius, and William Thomson, Lord Kelvin. Two important developments were required for this formulation.

One was the recognition of the importance of quantity and measurement in the operations of motors, which revealed that no mechanical process was entirely reversible. It was not that feedback or recycling processes were impossible in principle, indeed, they were a crucial feature of many energy systems; but they would never allow as much work to be got out of a system as had been put in. For that to happen, one would have to have a time machine, or rather a timeless machine; and, from now on, that was no longer a possibility.

The second development was a change in the attitude to the possibility of perpetual motion in nature. Many schemes for creating perpetual motion depended upon divine or scriptural warrant, for it seemed self-evident that nature was both mechanical and perpetual. Peter Brentano's scheme for forcing the water which runs out of a tube back into it by the countervailing force of weights and levers finds its guarantee in passages of Genesis and Job which, he claims, 'beautifully illustrate that perfect method to cause Perpetual Motion, which nature unfolds' (Brentano 1830: 13). Others before him had looked to the apparent proof of perpetual motion provided by nature. In his *Dissertation on Effervescence and Fermentation* (1694), the Swiss mathematician John Bernoulli articulated a common line of reasoning:

> Does not Nature herself (that is never said to operate otherwise than according to mechanical laws) show that Perpetual Motion is possible? What, to recall this example alone, is the perennial flux of rivers and seas but Perpetual Motion? And is not all this performed mechanically? So you must admit that it does not exceed mechanical laws and is not impossible. What then prohibits us from following Nature in this, in order to imitate her perfectly? (Bernoulli 1742: 1.41-2; my translation)

So the second development, a far-reaching, even shocking consequence of the first, was the recognition that the reason it was not possible to build a perpetual motion machine on natural principles was not because it was impious, or because of any limit to human capacities, but because there could be no perpetual motion in nature either. Perpetual motion, or maximal efficiency, was in a sense the defining horizon of possibility of all machines, but the impossibility of reaching this condition was what characterised all machines, including the machine beyond machinery of nature itself.

Henry Dircks's *Perpetuum Mobile* (1870), an historical review of perpetual motion schemes, leaves no room for ambiguity:

> The present century is rife in the reproduction of patented blundering, serving only to prove the ignorance and mental imbecility of a certain class of infatuated, would-be inventors, whom no history can teach, no instruction reform, nor any amount of mechanical mishaps persuade to abandon their folly. Anything more pitiable than such patented abortions it is almost impossible to conceive; while at the same time one would think that the perfect ease with which inventors might experimentally prove the fallacies of such model monstrosities, would suffice to enlighten them, without their incurring patent expenses to secure a right to the possession of such nothings. (Dircks 1870: xii)

Dircks does however allow himself some teasing reflections on the fact that the obstinacy of these infatuations seems in itself to demonstrate the very principle of infinite recurrence that has proved so hard to realise in practice:

> It is an admitted fact that whatever has been may again come to pass; so that even lost discoveries and inventions may, in the course of time, come to light again; in short, that human ingenuity is ever reproducing itself. It is evident, therefore, that while lacking information in regard to the past, the inventor, by following in the track of his predecessors, does little more than take up a circuitous route ending where it began. It is this very want of progress that marks every successive attempt to destroy the inertia of matter through the medium of mechanical arrangements put together in defiance of the laws of gravitation. (xi-xii)

Dirks even informs his reader, with amused resignation, of a number of inventors who, far from being discouraged by the absurdities collected in the first edition of his *Perpetuum Mobile*, were spurred by it to further deluded quests of their own. One kept a copy under his pillow so as to be able to fall upon it without delay first thing in the morning (xvi). Another bought the book for the support he expected it to provide for his own settled conviction as to the impossibility of perpetual motion, only to become addicted to his own investigations into the perpetuist possibilities of the unevenly weighted wheel, before abandoning his enquiry, 'finding that it proved too exciting for his brain, causing him to talk on the subject in his sleep' (xviii). The quest for perpetual motion became one of the signature forms of monomania reported on by nineteenth-century medical writers (Duffy 2010).

One of the wilder adherents of perpetual motion was William Martin, who in 1821 announced 'a Natural Philosophy, founded on the Principle of a Perpetual Motion, and the great Cause of such Perpetual Motion'. Martin bases his philosophy on Genesis 2.7. 'And the Lord God formed man of the dust of the ground and breathed into his nostrils the breath of life, and man became a living soul'. Martin draws from this verse the principle 'that AIR is the real Cause of Perpetual Motion' (Martin 1821: 12). His book uses this discovery to argue, among other disconnected things, that the sun is many times smaller than estimated by Newton. Martin tells us that the truth of this secret animating principle of the universe

> has hitherto been slightly noticed, or rather indeed, entirely overlooked by our great Philosophers, from whom (as vainly preferring their own imaginations, and too much neglecting the study of GOD's Holy Word) the Great Almighty has thought proper to conceal this important Discovery, until mankind had arrived at a more perfect state; and, to make it a matter of greater admiration, to illuminate, by His Divine Assistance, the mind of a poor simple individual (favoured by education only in a small degree) with light sufficient to make this great Discovery. (10-11)

Martin's book does describe a bizarre perpetual motion device, consisting of a pendulum precisely 39 2/10 inches long actuated in some mysterious way by a flow of air conducted through a ventilation shaft from the outside to the inside of a house (66-7), as well as devices for preventing dry rot, extinguishing fire at sea, and ventilating coal-pits. But really the intention of his book is to reveal a whole delirious cosmic system.

So we may say that it is not just the object of the dream, but the dream of perpetuity itself, that is perpetually recurrent in the *perpetuum mobile*. Perhaps that temptation is perpetual, or at least enduring, precisely because the temporality of technical development seems to approach the goal of perpetual motion asymptotically. Oswald Spengler sees the dream of perpetual motion as marking the end of the reign of Faustian *techne*:

> This is the significance of the *perpetuum mobile* dreamed of by those strange Dominicans like Petrus Peregrinus, which would wrest the almightiness from God. Again and again they succumbed to this ambition; they forced this secret out of God in order

themselves to be God. They listened for the laws of the cosmic
pulse in order to overpower it. And so they created the *idea of
the machine* as a small cosmos obeying the will of man alone.
(Spengler 1926: 2.502)

But Spengler sees a reversal at the point at which man's 'outward- and upward-straining life-feeling', the 'ineffable longing [that] tempts him to indefinable horizons', was perfected by means of an acceleration of communications technologies such that 'tiny man moves as unlimited monarch' (2.503). For suddenly

> these machines become in their forms less and ever less human, more ascetic, mystic, esoteric. They weave the earth over with an infinite web of subtle forces, currents, and tensions. Their bodies become ever more and more immaterial, ever less noisy. The wheels, rollers, and levers are vocal no more. All that matters withdraws itself into the interior. Man has felt the machine to be devilish, and rightly. It signifies in the eyes of the believer the deposition of God. It delivers sacred Causality over to man and by him, with a sort of foreseeing omniscience is set in motion, silent and irresistible. (2.503-4)

We may perhaps identify the modern devil devised by James Clerk Maxwell, in his information-based version of the *perpetuum mobile,* with the medieval devils who hovered in the vicinity of medieval automaton-makers. It is as though Spengler sees a world in which a kind of perpetual motion has been achieved, and as a result has lost all aspiration or propulsiveness. In his *Man and Technics* (1932), Spengler enlarges on this picture of the lifeless dominion of the machine:

> All things organic are dying in the grip of organization. An artificial world is permeating and poisoning the natural. The Civilization itself has become a machine that does, or tries to do, everything in mechanical fashion. We think only in horse-power now; we cannot look at a waterfall without mentally turning it into electric power; we cannot survey a countryside full of pasturing cattle without thinking of its exploitation as a source of meat-supply; we cannot look at the beautiful old handwork of an unspoilt primitive people without wishing to replace it by a modern technical process. Our technical thinking must have its actualization, sensible or senseless. The luxury of the machine is the consequence of a necessity of thought. In last analysis, the machine is a symbol, like its secret ideal, perpetual motion – a spiritual and intellectual, but no vital necessity. (Spengler 1932: 94)

The perpetual motion machine seems to conjoin and feed into one another two kinds of work that have traditionally been separated: the kinetic work performed by muscles and machines, and the metaphorical or imaginary work performed in imagining. This recapitulates a distinction between energy and what an Aristotelean tradition called *energeia*. Somewhat surprisingly, the first appearance in English of this word is in a passage from Philip Sidney's *Apology for Poetry*, where it is used to signify force or vigour in the language of poetic seduction – what in language seems to pursue rather than simply signify the work of love:

> But truly, many of such writings as come under the banner of unresistible love; if I were a mistress, would never persuade me they were in love; so coldly they apply fiery speeches, as men that had rather read lovers' writings (and so caught up certain swelling phrases which hang together like a man which once told me the wind was at north-west and by south, because he would be sure to name winds enough) than that in truth they feel those passions, which easily (as I think) may be betrayed by that same forcibleness, or *energia* (as the Greeks call it) of the writer. (Sidney 2002: 113)

Sidney is here referring to the discussion in Aristotle's *Rhetoric* III.xi of the means whereby metaphors make us see things, which, for Aristotle, depends on using expressions that represent things in a state of activity. The word *energeia* that Aristotle coins for this, and which he tends to use interchangeably with his other coinage *entelechy*, indicates the work of making-actual, or the condition of being at work. Sidney's usage is revived in Humboldt's concept of *energeia* in his *On Language* (1836), where it is introduced to express the relation between structure and performance in language, a distinction that would be re-formalised as that between *langue* and *parole* by Ferdinand de Saussure:

> *Language*, regarded in its real nature, is an enduring thing, and at every moment a *transitory* one. Even its maintenance by writing is always just an incomplete, mummy-like preservation, only needed again in attempting thereby to picture the living utterance. In itself it is no product (*Ergon*), but an activity (*Energeia*). Its true definition can therefore only be a genetic one. For it is the ever-repeated *mental labour* of making the *articulated* sound capable of expressing *thought*. In a direct and strict sense, this is the definition of *speech* on any occasion; in its true and essential meaning, however, we can also regard, as it were, only the totality of this speaking as the language. [...] It alone must in general always be thought of as the true and primary, in all investigations which are to penetrate into the living essentiality

of language. The break-up into words and rules is only a dead makeshift of scientific analysis. (Humboldt 1999: 49)

All perpetual motion machines must take some kind of circular form, since they feed back all the energy required for their working into that working. They must both be in a state of as it were unbalanced equilibrium, never accelerating nor retarding, while they must also deny the approach to static equilibrium that is the tendency of all machines. A good image of this paradox is the idea for a perpetual motion machine proposed in a thirteenth-century notebook by Villard de Honnecourt which suggests that a wheel might be put into permanent rotation if it were hung with an odd number of weights on one side and an even number on the other, because it will always be out of balance (Bowie 1959: 134). Leonardo's analysis of this idea disclosed that the wheel would be bound to reach a state of static rather than dynamic equilibrium, with one weight in the middle and two on each side, from which it could not be shifted without the addition of some new energy (Olshin 2009: 2, 6-7). The relation between the completed work of language, and the perpetual coming into being of language, seems equivalent to this kind of permanent state of becoming that never settles into any permanent condition of being.

Though widely accepted among physicists for most of the second half of the nineteenth century, following the formulations of Carnot, Clausius, and Thomson, the theoretical basis for the second law of thermodynamics was not satisfactorily explicated until Ludwig Boltzmann explained it in terms of probability. Boltzmann showed that dynamic systems always move from an ordered to a disordered state – where an ordered state might mean one in which work could be performed because of some continuing differential, of weight, heat, or anything else. So a hot liquid brought together with a cold will spontaneously move towards an equilibrium state, never one in which the hot water gets hotter still and the cold colder still. The reason for this is that there are just many more ways in which the former condition can arise than ways in which the latter can, just as there are many more ways in which a pack of cards can be disordered than ways in which it can be ordered into suits and values. But this explanation actually meant that the Second Law was no longer a law, so much as a likelihood, even though it was one that was, for all practical purposes, a racing certainty. Boltzmann opened the door, if only by a chink, to new ways of thinking about perpetual motion. Now, it was a matter not of manipulating relationships of force, but rather of playing with probabilities.

The new way of thinking was assisted by the second feature of Boltzmann's work, namely the understanding of energy in terms of information, where the emphasis of that word, as in its earliest theological uses, is in the idea not just of communicating knowledge, but also in the process imparting a form, shape, or order to what has previously been formless. Thomas Browne used the word in this sense in his *Religio Medici* (1642) to signify the Creation, writing that 'God being all things is contrary unto nothing out of which were made all things, and so nothing became something, and Omneity informed Nullity into an essence' (Browne 1977: 105). As many have noted, something decisive happens to the idea of energy during the 1940s, with the work of Norbert Wiener, Claude Shannon, and others starting to hold out the possibility of treating energy and information in the same terms. Heat, force, and energy were now to be understood as forms of arrangement. Even more remarkably, things that had previously been seen as entirely intractable to mechanical and quantitative explanations, like language, literature, media, and signs, were suddenly made comprehensible in these terms.

Among the two most important figures in popularising these notions are Michel Serres and Thomas Pynchon. In his essay 'The Origin of Language', Serres announces a 'new organon which maintains the advantage of being at the same time a physics of energy and a theory of signals' (Serres 1982: 81). Two different kinds of argument are offered in Serres's essay. According to the first, information theory makes what Serres calls the 'soft' – ideas, forms, semiotic structures – amenable to the same kinds of understanding and explanation as are applied to the 'hard' – physical objects and processes. This need not, of course, imply that the orders of physics and of semiotics are in fact the same thing, only that they may be modelled in similar ways. But Serres goes much further than this argument through resemblance. A living organism may be thought of as inhabiting a space between a physical system and a system of symbolic communication: a system for distributing energies, and a system for transmitting messages. Serres borrows from fluid dynamics to image the emergence of stable form from dynamic process:

> the organism is a barrier of braided links that leaks like a wicker basket but can still function as a dam. Better yet, it is the quasi-stable turbulence that a flow produces, the eddy closed upon itself for an instant, which finds its balance in the middle of the current and appears to move upstream, but is in fact undone by the flow and re-formed elsewhere. (Serres 1982, 75)

This seems to give warrant for thinking of the body as a kind of negentropic machine, one which resists the drift into disorder, through a series of 'black-boxing' encapsulations, resulting in 'an extraordinarily complex system that creates language from information and noise, with as many mediations as there are integrating levels' (82). Serres is able to conclude that

> [n]othing distinguishes me ontologically from a crystal, a plant, an animal, or the order of the world; we are drifting together toward the noise and the black depths of the universe, and our diverse systemic complexions are flowing up the entropic stream, toward the solar origin, itself adrift. (83)

In a more recent essay on Virginia Woolf's *To the Lighthouse*, Serres seems to offer the strongest possible version of this claim. The central section of the novel, 'Time Passes', evokes a house that is falling into disorder through lack of human inhabitation. The return of the family to the house suddenly arrests this entropic slide, giving order and purpose to it once more. Serres sees no significant difference at all between the restoration of order effected by the cleaners who energetically set about dusting and polishing the neglected house, and the work of memory, perception, and anticipation which restores human meaning to the house, whether at the level of the characters in the fiction, especially Lily Briscoe, or at the level of the narrative itself:

> We work and die hard; but we think and perceive softly. Ill winds blow through the world, hard and destructive, but also soft breaths that resemble spirit and intelligence. This very softness, this software, then becomes as real as the material itself, as real as the hard.
>
> Immense revolution in the sciences and in labour. (Serres 2008: 116)

Serres asks himself the hard question about how the hard is affected by the soft, and answers only with the poetic assertion one might think it his task to explain: '*esse est percipi*. To perceive beings fills them with being. Simply by their perceptions, human beings – living things? – in this case women, fabricate negentropy, produce information, and thereby oppose the irreversible degradation of things' (124). The analogy between physics and symbols seems to permit the idea of a single process, governing all, and allowing one to write equations that can encompass soft and hard and make them commensurate. Instead of a perpetual machine, we seem to have a perceptual machine with equivalent powers. There may be huge differences of scale in the energy budgets that one would draw up for the writing of a symphony and the making and digestion of a stew, but they are not on this account fundamentally different processes. This is not just because there is a kind of energy budget to be drawn up in every act of imagining or reasoning – something of the hard in every exercise of the soft – but also because of the apprehension that, at the heart of matter, and at a level that it proves impossible to get behind or underneath, there seems to be organisation, code, information, indeed, there seems to be nothing but these:

> Our classical sciences calculate relations of power and energy, forces, and causes of movements. In so doing, they remain on the entropic scale. But today these sciences also go on to describe in all things the storage, emission, reception, and processing of information. They teach us that the soft penetrates the hard. (127)

One can acknowledge readily enough that matter is in fact organised at the smallest conceivable or calculable scales, and that the workings of matter are understood better as exchanges of information than as the application of force to inert and insentient stuff. One can also acknowledge that a lighthouse, which Serres takes as an image of a perceiving gaze, can work on the world and transform it, in the same way as a dredging-barge or dreadnought cruiser can. But this does not explain the mechanism of mediation between these scales, which is just where what we understand as mechanical difficulty has always arisen.

What might it mean, for example, to try to draw up the energy account for the very process of which you are seeing one outcome, in these words? Were I incising these words in clay or stone, there would be a large gap between the process of conceiving the words which would embody my meaning and their execution. I would need to decide what needed to be written, and suspend the operations of thought while I – or perhaps my inscribing slave or stenographer – put them into the world, at a palpable cost in physical energy. But I have been typing these words – indeed, the process is so softly accommodated to the eddyings of my thought that the word typing itself seems too close to the world of metal and the foundry, implying too much a work of casting. I have time, not just in the forming of a sentence, but in the forming of an individual word, to rethink it, and, of course, my word-processing software allows me at every stage to delete and undelete, so that the writing process is riddled with temporality. More and more, it seems as though whatever thinking might be, it is no longer a matter of sending an order to be printed – 'Execute', as the old computers used sometimes to call the Return key – but is rather mingled with the technologies of writing. Soft media mean that mediation both vanishes – there is no delay between conception and inscription – and enlarges, to encompass the entire process from beginning to end, so that the entire process occurs in the middle. In our world, the work of forming words and other kinds of inscriptions seems to involve very little labour that can be called hard at all. Making a poem, picture, or proclamation used to require a large amount of energy; in our world the energy required is almost negligible, which indeed seems to make the hard increasingly indistinguishable from the soft.

And yet not. The very cheapness of such media means that they tend to multiply and indeed that they need to multiply in order to become even cheaper. Were I Roger Bacon working away in Oxford in 1290 at the unique example in the world of the magical writing machine I am now employing, conveyed to him by means of time travel or donated by a passing extraterrestrial, the cost in time and labour of writing would indeed have seemed to have shrunk to the point where it was negligible, and where he was getting something for as near as anything nothing at all, that is, a machine that, as long as his thinking continued, would instantiate his thought magically and without loss in the world. But, in fact, I can only write in this fashion, and you can only read these words, transmitted as though from one telepathic angel to another, as a result of a huge scaling up, whereby billions of individual people casually employ computing machines powerful enough to calculate a passage to the moon to check Facebook several times an hour. At these scales, remarkable things happen.

Serres suggests one thermodynamic reading of the development of culture among homoiothermal organisms like humans – that is, organisms that have to maintain their body temperatures artificially, rather than allowing them to fluctuate with the thermal state of the world. Bees in a hive generate and distribute warmth by fluttering their wings. Human beings do it through social and communicative means:

> The homoiothermal organism generates the need for communication. It is, in energy or thermal needs, analogous to what will be common speech, in terms of signals and information. I imagine that one of the first forms of behaviour, like one of the first signals, may be reduced to this: 'keep me warm'. The homoiothermal organism initiates touch and contact, erotic communication, and language. (Serres 1982: 76 n.6)

Peter Sloterdijk similarly evokes a 'global civilization greenhouse' as the image of an immunological culture that keeps us safe from chilling exposure to the world outside, even as the creation of that very immune system, whether in blankets of warmth formed from the combustion of fossil fuels, or by means of an 'electronic medial skin', entangles us in new 'thermo-political paradoxes' (Sloterdijk 2011: 25), meaning that we have to find ways of keeping cool our ways of keeping ourselves warm. For we have known for some considerable time that our problem is not so much how to generate heat to protect us from a cold world, but rather how to prevent our actions and communications, along with our actions that increasingly take the form of communications – from producing too much heat. Our communications in fact generate a large and raw thermodynamic cost, of heating. Combustion engines produce large amounts of greenhouse gases, but there is also a huge and growing demand for means to keep computing equipment from overheating. The almost ubiquitous sound of the modern office is the sound of the internal fans humming inside computers to cool them down. Instead of the thermocommunicative solicitation between organic creatures – 'Keep me warm' – there is the imperious demand of our machines for expensive, heat-producing, and entropic auxiliary machines to keep them cool. Instead of the satanic mill, there is the server farm, often placed in cold locations like Iceland to give the cooling mechanisms a head-start. So there is a net cost in excreted and irretrievable heat for every degree of coolness achieved or maintained.

No more intransigent example can be imagined of the operations of the Second Law, which is often represented in terms of the tendency for hot things spontaneously to cool rather than the reverse. This is true only where cooling represents the entropic movement to equilibrium. But in a heating world, in a world where there is always a cost in unusable heat for every mechanical or electronic operation, the drive to equilibrium takes the form of apparently spontaneous warming rather than cooling. Refrigeration came late on the scene in historical time, and employs methods that themselves seem like a by-product of perpetual motion devices, precisely because of the force of the Second Law, which makes it much harder to produce artificial cooling, which usually involves a movement away from equilibrium, than artificial heating. In a cool, mechanical world, human beings would usually, at 37C, be the hottest things in their environment, aside from the fire around which they might be clustering, hotter than rock, metal, wood, water, plants, and many other animals. The leading problem in the period prior to global warming, in many of the environments which human beings colonised, may well have been how to prevent heat-loss.

Modernity may be measured by the way in which humans begin to cohabit with technical objects that require to be operated at much higher temperatures than previously. The temperature in Basra on 21st July 2016 was one of the highest ever recorded, at 54C: such extreme temperatures are being recorded much more commonly. The striking thing about this is how much above human body temperature it is. We all know how to get warm when it is cold; our desperation in conditions of extreme heat comes from the fact that it is so much harder to cool something hot (most notably a sweltering body) than to heat up something cold, since the mechanical actions required to produce cold all produce heat as a by-product. Heat is universal and easily procured, cold on demand is rare, precious, and luxurious. Any peasant can warm himself over a pile of smouldering turf; until recently, only an empress could command her sweating slaves to serve her sorbet in midsummer. We give the word 'heat' both to the measure of an amount of energetic motion, and to the positive quality we imagine heat to be. It makes sense to think of 'more' or 'less' heat where it does not seem to make sense to speak of quantities of coolness or non-heat, any more than it once did to measure levity, or non-heaviness (Connor 2009). For centuries, it was thought that each creature was allotted a certain portion of vital heat that was identical with their life. But, in a world of rising and perhaps now uncontrollable temperatures, the abstract idea of coolness has indeed come to take on a positive value. Most importantly, it requires a machinery based on the principle, not of self-perpetuation, but of self-limitation, through active monitoring. For Marshall McLuhan, a cool medium is one of implication, in which we have ourselves to supply much of the information – novels, or radio, for example, as opposed to explicit and high-definition media such as film and TV (McLuhan 1994: 22). As our media have heated up, in tandem with global temperatures, our imaginary machines have perhaps begin to cool down, precisely because they require so much from us of projection and systematic self-monitoring.

The 'bachelor machines' defined by Michel Carrouges share certain characteristics with perpetual machines, especially in their serenely sequestered self-reference. Contemporary with these visionary, immaculately sterile apparatuses is the project given witness in *The Perpetual Motion Machine* (2011), a book first published in 1910 by Paul Scheerbart, a prolific writer of visionary and speculative fictions. The book affects to document the process which Scheerbart himself undertook to invent a perpetual motion machine, of the most traditional kind, namely an 'overbalancing' machine. The book announces his success in a preface, but in fact the documentation of his toilings and blunders, as pitiable as they are comic, begins to take the place of the machine, even, in a way, to become its apotheosis. The story of the machine and the machine of story become identical.

Scheerbart's work anticipates the growing sense of the parallel between machines and the images and texts that mediate them that grew through the century, especially following the development of cybernetics and information theory in the years following the Second World War. Thomas Pynchon paid particularly close attention in his work to the entanglements of mechanical and mediatic form, sometimes through the mediation of imaginary perpetual motion machines. *The Crying of Lot 49* features a machine invented by a physicist called John Nefastis. The point about this machine seems to be that there is no effort to make it even appear as though it functions as a machine at all. It consists simply of a photograph of James Clerk Maxwell on the top of a box out of which come two pistons attached to a crankshaft and flywheel. Another engineer called Stanley Koteks gives Oedipa Maas, the novel's heroine, an account of the workings of the machine, which is really just a paraphrase of the usual account of Maxwell's thought experiment about the demon who might simply sort molecules into hot and cold, thereby creating the capacity for work without actually putting any work into the system. Responding to Oedipa's scepticism about the idea that sorting is not work, Koteks insists that mental work is not work in the physical sense, which is what allows the machine to violate the Second Law of Thermodynamics, 'getting something for nothing, causing perpetual motion' (Pynchon 1982: 62).

Later in the novel, Oedipa meets Nefastis and is given a chance to test whether she is the kind of 'sensitive' who is able to communicate with the demon and cause it to concentrate molecules sufficiently to move one or other of the machine's two pistons. Nefastis describes entropy as a metaphor which connects the worlds of physics to information flow: 'The Machine uses both. The Demon makes the metaphor not only verbally graceful, but also objectively true' (77). The Demon and the machine it is imagined as actuating (though the Demon really is the machine in fact) are a metaphor for a metaphor that might work as an actuality: that should, like the poem of Archibald MacLeish's 'Ars Poetica', not just mean, but be (MacLeish 1985: 107). Oedipa strains to make the machine work, thinking 'If you are there, whatever you are, show yourself to me. I need you, show yourself' (Pynchon

1982: 79). The machine is an imaginary concentration, not just of Maxwell's dramatised conundrum, but also of the mysterious convergence of ideas, communications, and actualities that Oedipa Maas is trying to interpret in the world around her. It is, therefore, something like a metaphorical *mis-en-abîme* for the whole of *The Crying of Lot 49* itself. At the end of the novel, Oedipa is left suspended between the two alternatives that it has constructed, that America is just America, and that she has in fact stumbled on a secret conspiracy, connecting everything and making the whole world at once a dissimulation and the single connected truth of that dissimulation. Nothing she is able to do or think allows her to escape from this binarism of meaninglessness or meaning, with every middle excluded. As she waits for an auction to begin that may or may not deliver her the secret, there is a reminder of her unsuccessful attempt to communicate with, or become the demon in, Nefastis's machine: 'She stood in a patch of sun, among brilliant rising and falling points of dust, trying to get a little warm' (137), the phrase 'getting warm' nicely conjoining thermodynamics and the approach to hidden knowledge. By this point, the imaginary machine is a metaphor for the machinations of the whole novel, which amounts to a machine for deciding if it is itself a machine or just a metaphor.

The most important point about the Nefastis machine is that, although it seems to work as a classical form of black box, it in fact figures an entire system of connections and transmissions, including, but potentially extending far beyond, the novel itself, of which it is both a component and a totalising image. The machinery involved is W.A.S.T.E., a centuries-old alternative postal system, formed from movements of information rather than energy, occurring across space and time rather than within some localised system of mechanical relations. Its perpetual mobilisation occurs across and among, rather than within and between.

Perpetual motion machines began to change their character around the turn of the twentieth century, just at the point at which the gloom at the thought of the heat death of a finite universe, as embodied in the glimpses of the end of the world contained in H.G. Wells's *The Time Machine*, and having their cultural correlatives in fears about degeneration, seemed to be at its deepest (Mousoutzanis 2014: 47-90). From the beginning of the twentieth century, the vision of steadily degrading energy gave way to more expansive visions based not on the closed system of the classical heat engine, in which the law of inexorably increasing entropy applied, but on ever larger and more open systems: instead of the totalisable 'machine', a generalised mechanism. What was at issue was not the perpetual motion of a machine sequestered from the external workings of nature, but a machine made perpetual by participation.

In response to the growing authority of the Second Law, many mechanical speculators began to seek, not for a way to make a machine capable of recycling without loss all the energy required to operate it, but for a source of energy that was so close to being infinite that it would deliver something that approximated to perpetual motion. One of the most sustained of these imaginative, or rather perhaps imaginary, engineering endeavours was that of John Worrell Keely. Keely claimed to have found a way to disintegrate substances like air or water in such a way as to capture the 'etheric vapour', later known by many other names, which held together the atoms of such substances, and employ it for mechanical purposes. The essential principle on which Keely's apparatus depended was that of the power of sympathetic resonance. Most of his devices used sound vibrations to unlock the prodigious powers of etheric vibration within matter. Nearly all of them are now lost, but examples and photographs do survive.

Most hoax or hopeful energy devices depend on being able to be demonstrated in some more or less plausible physical form. Keely's endeavours from the early 1870s until his death in 1898 are distinguished by the fact that he toiled so ceaselessly at the production of his various machines for freeing and harnessing etheric force, creating and experimenting with hundreds of prototypes, many of which were destroyed in explosions, at a cost of thousands of dollars (Paijmans 2004: 165-6). Indeed, the distinctive feature of Keely's speculations is that they resulted not in one machine, but in hundreds, bearing names like the Multiplicator, the Liberator, the Transmitter, the Compound Disintegrator, and their hyphenation-highlight, the Hydro-Pneumatic-Pulsating-Vacuo Engine, as though to demonstrate something like a self-generating and self-perpetuating power in the mechanical imagination itself. After Keely's death, his laboratory was found to be provided with elaborate hidden pipework and mechanical belts and switches, suggesting that many of the effects he procured relied upon the concealed operations of compressed air (Hering 1924: 96).

Keely was financed by the Keely Motor Company, which was established in 1874 on the strength of some seemingly persuasive practical demonstrations, and would attract capital of $5 million (Paijmans 2004: 25-6). In later years, Keely was supported by a wealthy Philadelphia heiress, Clara Jessup Bloomfield Moore, who also wrote loyal and laudatory accounts of his discoveries. One would have thought that, with all the time and labour Keely expended, and with all the resources he was able to marshal, the chances of an investigator of reasonable perceptiveness and intelligence making some practical discovery by accident must have been quite high; so it is surprising that no such discoveries of any kind were made.

Inevitably, as Keely's researches and speculations grew ever more encompassing, he began to suggest that his etheric energy was continuous with and so perhaps governable by the power of thought alone, with the mind as 'interface or controlling mechanism between this material aspect of existence and the invisible cosmic life-force, the sea of light and willpower, and the emanations of the one supreme being' (Paijmans 2004: 182). Clara Bloomfield Moore explained that 'the only true medium which exists in nature is the sympathetic flow emanating from the normal human brain, governing correctly the graduating and setting-up of the true sympathetic vibratory positions in machinery necessary to success' (Bloomfield Moore 1893: 50-1). This is perhaps Keely's version of the reflexivity identified by Freud in his characterisation of all magical thinking as governed by an 'omnipotence of thoughts' which is simultaneously desired and feared (Freud 1953-74: 13.84-5). The work performed by Keely's elaborate and proliferating energy apparatuses, along with the tireless work of discourse that explicated them, was not kinetic but semiotic – a noetico-poetic quasi-physics. It existed to figure the operations of a mind capable of capturing and governing its own fantasies of mechanism and mechanising its own powers of fantasy.

The discovery of radioactivity in the 1890s meant that 'vibration' was overtaken by 'radiation' as the form of energy most invested by fantasy. Early in the twentieth century, Frederick Soddy's work on radioactivity was instrumental in prompting a host of the radiation-machine fantasies considered in Chapter Six. In *The Interpretation of Radium* (1909), he wrote that

> the driving power of the machinery of the modern world is often mysterious, but the laws of energy state that nothing goes by itself, and our experience, in spite of all the perpetual motion machines which inventors have claimed to have constructed, bore this doctrine out, until we came face to face with radium. Nothing goes by itself in Nature, except apparently radium and the radio-active substances. That is why, in radioactivity, science has broken fundamentally new ground. (Soddy 1909: 29)

Suggesting of radium that, aloof and indifferent to its environment, '[i]t seems to claim lineage with the worlds beyond us, fed with the same inexhaustible fires, urged by the same uncontrollable mechanicism which keeps the great suns alight in the heavens' (38), Soddy suggested that radioactive reactions provided, 'in the larger laboratory of Nature, an example of practical perpetual motion on the grandest and most majestic scale' (33). On a more earthly scale, Soddy was impressed by William Crookes's spinthariscope, a device which displayed as scintillations the collision with a zinc sulphide screen of alpha-particles emitted by a very tiny piece of radium – so tiny that that device could be purchased for a few shillings from an optician. Soddy marvels that the screen will wear out before the stream of particles shows any sign of diminishing, and indeed that '[t]he owner of the instrument will pass away, his heirs and successors, and even his race, will probably have been forgotten before the radium shows any appreciable sign of exhaustion' (61). He opens up the possibility of new kinds of radiant mechanics, centred on the workings of the atom, no longer conceived as an elementary thing, but rather as 'an almost infinitely complex piece of mechanism' (217).

However, though he remained sure that the derivation of energy from radioactive processes would be important, Soddy devoted the second half of his life to a campaign for the chastening of fantasy in economics, which he saw as 'concerned with the *interaction* with the middle world of life of these two end worlds of physics and mind' (Soddy 1922: 6), by linking it to the laws of thermodynamics. That is, he aimed to resist and repudiate the idea of economy as 'a presumed perpetual motion machine' (Daly 1986: 202-6). Soddy criticised what he called 'Ultra-Materialism', which he described as 'the attempt to derive the whole of the phenomena of life by continuous evolution from the inanimate world' (Soddy 1922: 7). He wrote that

> I cannot conceive of inanimate mechanism, obeying the laws of probability, by any continued series of successive steps developing the powers of choice and reproduction any more than I can envisage any increase in the complexity of an engine resulting in the production of the 'engine-driver' and the power of its reproducing itself. (7)

Strangely, though, the example that he gives to distinguish dead from living matter is the difference between 'Niagara Falls thirty years ago and now' (7), which is to be explained by 'the operations of intelligence, as typified in their most rudimentary form by Clerk-Maxwell's conception of the "sorting-demon"' (7). This may tell us nothing about where intelligence might have come from, but it does seem to imply that matter is not just added to, but somehow changed, when subject to sorting.

Nevertheless, Soddy recommended for economic thinking a much more traditionally mechanistic understanding:

> Let us now leave generalities and concentrate upon the question as to what precisely humdrum mechanical science can contribute to economics. It insists primarily on the fact that life derives the whole of its physical energy or power, not from anything self-contained in living matter, and still less from an external deity, but solely from the inanimate world. It is dependent for all the necessities of its physical continuance primarily upon the principles of the steam-engine. The principles and ethics of human law and convention must not run counter to those of thermodynamics. (7)

This means paying attention to the problems of energy, which are the most important kinds of problem for men, who are 'no different from any kind of heat engine' (7). Human beings, like other entities in nature, consume energy, which ultimately is only some form of sunshine, whether direct (solar radiation) or sedimented (coal). To say that this revenue of energy is 'consumed by the living engine in its life' is not at all to say that it is destroyed, for that would contradict the law of conservation of energy, but rather that 'it finds its way into the great energy sink, the ocean of heat energy uniform with the surroundings, and is incapable of any further transformation [...] it is useless' (7).

There are some unsavoury elements in Soddy's writing. He argued that '[W]ealth is a flow and cannot be saved' (58). To convert wealth to debt is to substitute a system of symbols for the thermodynamic fact that energy flow is irreversible. This means that a system which is dependent upon the derivation of unlimited revenues from interest is a kind of fantasy perpetual motion machine. Wretchedly, Soddy gives this fantasy, or its deleteriously parasitic effects, the name of 'the Jew':

> The capitalist wishes to have it both ways, to be regarded as a public benefactor because he spends his wealth, not in drinking himself to death, but in enterprises designed to increase the revenue. If he did this he would indeed be a public benefactor. But the community having spent his wealth, as regards himself he expects it all back in due course with interest on the loan. The consequences of his abstinence are that civilisation has got inextricably 'into the hands of the Jews.' (25)

Predictably, Ezra Pound found aspects of Soddy's thought attractive, quoting him approvingly in *Guide to Kulchur* (Pound 1970: 245-6). Nevertheless, Soddy's energy-economics have come to exert considerable traction in recent years, especially given the co-presence of ecological pressures in economic thinking, and the recent experience of extreme and catastrophic financial instability based on unmanageable debt.

In 1900, Nikola Tesla published 'The Problem of Increasing Human Energy' in the *Century Illustrated Magazine*, in which he records his speculations about the history and future of humanity, considered in the largest possible terms, as 'a mass moved by a force' (Tesla 1900: 192-3), where the mass is the total number of human beings, and the force includes all the forms of human invention and motivation. Tesla sees the problem of maximising energy as a mechanical one, the solution of which would need to operate on a very large scale, to take account of all the complex and interdependent forms of mechanical energy that are a feature of modern human life. Our dependence on these forms of mechanical energy produces a sense of anxiety: '[W]hen there is an accidental stoppage of the machinery, when the city is snowbound, or the life sustaining movement otherwise temporarily arrested, we are affrighted to realize how impossible it would be for us to live the life we live without motive power' (191-2).

The solution to this problem of ever-increasing technological dependence is an extension of technical thinking to encompass even more of human life. Tesla is encouraged in this view by his reflexive conviction of his own mechanical condition. He describes the difficulty he experienced as a boy from the 'the appearance of images which, by their persistence, marred the vision of real objects and interfered with thought' (184). An intense period of self-examination convinced him that all of the images and thoughts he experienced could be traced back to antecedent thoughts and images, which therefore formed a chain of mechanically-determined consequence:

> searching, observing, and verifying continuously, year by year, I have, by every thought and every act of mine, demonstrated, and do so daily, to my absolute satisfaction, that I am an automaton endowed with power of movement, which merely responds to external stimuli beating upon my sense organs, and thinks and acts and moves accordingly. I remember only one or two cases in all my life in which I was unable to locate the first impression which prompted a movement or a thought, or even a dream. (184)

Convinced as he was, like the young man who said 'Damn' of the limerick, of his own automaticity, Tesla proposed a system of general human automation. His ideas for maximising human energy involve maximising population, through preventing waste of life in war, this to be achieved by the development of 'telautomatics', characterised as 'the art of controlling the movements and operations of distant automatons' (186). Automata with 'borrowed minds' (187) would fight humanity's wars by proxy.

However, the most important and influential proposal of Tesla's essay is the idea for a machine that would harvest energy from the 'ambient medium', that is, the air. In essence, it puts forward a scheme for creating a space of artificially reduced temperature into which heat from the air would naturally flow, with that flow being converted into electricity. This sounds pretty much like using a refrigerator as a heat-engine, and it seems obvious that one would always have to use at least as much energy to create the reduction in temperature as one would gain from the exchange. But, if it could be done with a tangible energy gain, one would then have produced what Tesla calls a 'self-acting engine' (201). Tesla offers some bold speculations as to how this might be done, which escape the strictures of Carnot and Kelvin that 'it is impossible for an inanimate mechanism or self-acting machine to cool a portion of the medium below the temperature of the surrounding, and operate by the heat abstracted' (201), because his machines draw energy from outside the system, and are thus not truly self-acting. One of them involves a thermopile, capable of generating thermoelectric voltage from temperature differences in different metals, extending from the earth to the upper atmospheres. There would be a flow of heat and also of electric current from the bottom to the top of the pile. If it were possible to join the two terminals of the thermopile, it could drive a motor which, theoretically 'would run on and on, until the media below would be cooled down to the temperature of the outer space' (201). Like Soddy's radiation energy, this is no longer strictly a perpetual motion machine, and so not a violation of thermodynamic laws, though, given the vastness of the reservoir of energy potentially able to be harnessed, it might as well be.

Tesla has a curious status. Like Edison, he became a fabular creature in literary speculation about human futures and possibilities. This is not at all because he was simply a fraud or crank, though he does seem to have had some very singular beliefs and preoccupations, for he was responsible for a number of extremely important and far-reaching inventions and innovations. But, going (even) further than Edison in his speculations, he became the vehicle of a powerful fantasy of technical wizardry, often associated with the idea that he had been victimised or ignored by an unimaginative or malicious scientific establishment. Tesla has made regular appearances in science fiction stories, films, and comic books. He became, that is, the embodiment of a powerful association between the idea of energy and the magical. Just as he viewed himself as a self-acting automaton, or 'borrowed mind', so, he has himself been automatised as the image of a technology-generating genius-energy.

In the last couple of decades, with the increasing need both to increase energy sources and to try to control the global warming that comes from combustion processes, Tesla's speculations about what has increasingly been called 'free energy' have become exciting to many. As with perpetual motion machines, there seems to be a limitless reservoir of fantasy about the possibility of devising mechanisms that derive energy for nothing. In 1996, an inventor called Dennis Lee demonstrated to an audience in Philadelphia what purported to be a heat pump working at room temperatures, deriving energy from the air. Such ideas, which have been documented by Donald A. Kelly (1991) and Eric Krieg (2002), persist in contemporary movements such as the Breakthrough Energy Movement, which mingles ideas like 'open system technologies' which are in principle plausible since they aim to harness energy from the environment rather than to manufacture it out of nothing, with frankly and frantically imaginary mechanisms. The host of the 2016 conference of the Breakthrough Energy Movement, which took place in Texas on May 28th and 29th 2016, was Vernon Roth, who claims to have learned of 'the essential interconnection between forces of energy and forces of nature' while studying Physics and Psychology as a young man. His website, Ancient Tek, markets a number of 'transformational elements' which use 'life force', which is another version of the special imaginary energy that has been believed in in different ways for many centuries and has taken pseudo-technical forms for the last two centuries:

> This is defined in ancient literature as Od, Aether, Orgone, and, in the body, as Chi and Prana. This same life force has been defined in today's scientific world as Aetheric Energy, Quantum

Energy, Dark Matter, and Superluminous light. It works not only at the physical level but the emotional and mental and spiritual levels as well. ('Ancient Tek' 2014)

It also makes available 'Spark of Life' water, which is capable of effecting remarkable cures and 'Glacier Milk', which, the website explains, is a powder found in glaciers.

> When the glaciers melt it [sic] drops out a white powder that after [sic] cleaned and purified is a natural ormus without the heavy metals that occur when ormus is processed with chemicals [...] has been used to block radiation, fight infections and poisonings, restore function to organs and has been implicated in facilitating remarkably long life spans. ('Glacier Milk' 2014)

The ecological imperative within energy science has encouraged this characteristic blending of technophobia with souped-up technofantasy. On the one hand, we are held to live in a world dominated by abstract mechanisation and the corporations which depend upon it for their profits. Proponents of free energy typically believe that geniuses like Nikola Tesla and their followers have had their work suppressed by a conspiracy of these corporate interests wielding the myth of the Second Law. On the other hand, there are the good forms of technology that have the potential to bring healing, peace and prosperity.

During the twentieth century, the dominion of machinery which had an obvious and quantifiable dependence on energy encouraged mechanical views, no longer of individual human beings, but of the processes that connected them. Attention turned away from Mettrie's homme-machine to the humanity machine. Machines enjoined a greater awareness of the importance of energy, which in its turn encouraged the view of collective relations as a kind of machinery. At certain times, notably during the mobilisation of resources and production for the purpose of war, this imaginary machine has been made actual.

Our perpetual motion machines are no longer those that can trick the universe into operating against its own principles. What we call a 'sustainable' mechanism is one that is able to derive energy from the perpetually renewed motions, of wind and wave, of the natural world. These no longer need to be imaginary machines, though many will still need to be imagined for the time being. But the action of imagining, and its energy-entanglement with media, makes it intimately part of the process it projects.

This is no longer a matter of dreaming up any kind of single, sealed-state imaginary device, which would perform a kind of magically counterfactual kind of mechanical work, but rather a total and all-inclusive machinery, which includes its own reckoning of the machinery in that machinery. It is not clear whether such machinery is to be thought of as an imaginary force or the force of imagination itself in climbing back up the entropic slope of time. It may be regarded as a kind of benign paranoia, in which the connectedness of systems seems to make it possible always to draw off more energy from some adjacent system.

This interfusion of technology and imagination has of course been intensified by the ever closer conjuncture of technology and media that is typified by the internet, the ever more actual, and ever more virtual, embodiment of Tesla's teleautomatics. The internet provides a fertile venue for imaginary technologies because more and more it seems itself to be one. Or rather, not exactly one at all. The internet is the embodiment of the idea of the universal meta-machine. This goes further than the 'universal machine' imagined by Turing, which would be capable of performing every imaginable task, since its principal function is beginning to be that of connecting up every other machine, actual, imaginary, and imaginable. Of course, the internet is far from performing this function in actual fact – access is expensive, connections are slow or patchy, and it seems certain that there will always be a more-or-less painful gap between what connected technologies, and technologies of connection, the two scarcely distinguishable any more, promise, and what they actually do, a gap that is simultaneously closed and held open by imaginary desire. The definite article of 'the internet' is a sign of the residual desire for the internet to be a machine, a delimited object set apart from other objects – as in the 'Over Logging' episode of *South Park* in 2008, in which 'the internet', imaged as a huge Linksys router, stops working until it is restored by Kyle by the homely expedient of turning it off and on to reboot it. The internet stands as an image of the total, all-comprehending machine that includes all other machines that, simply because there is no form in which it can be imagined as any sort of definite article, can only be imaginary.

This appears to be a perpetual motion machine that does not need to be monitored because it seems to be self-acting. But the very reason it can be relied upon to supply energy unceasingly is because its users supply it with energy in the very act of using it. Where the perpetual motion machines of the past were set aside from the rest of the thermodynamic universe, occupying a space of counterfactual exception in which they required no intervention or input because they ran on their own, the surrogate perpetual motion machine that is embodied in the fantasy of the internet, as image and enactment of the imaginary state of total communication of all earthly machineries whatsoever, requires constant input and monitoring, partly because it is increasingly the very mode of our own informational self-monitoring. Machines are supposed to work on their own, for us; the internet works on its own, but through us. It is perpetual not because it derives its energy from itself, but because all of its energy is derived from its users. A complete mediatic machinery of this kind is both hot and cold at once, in McLuhan's terms. It is hot because it operates with such fine-grained high definition, able to constitute the world with such fidelity as to substitute for it. But it is cool in that it runs entirely on belief, projection and supposition supplied by our input, which is also its output. What remains moot is whether the fantasy of the perpetual will indeed be sufficient to perpetuate itself.

Freud remarks in passing in a footnote to *Civilization and Its Discontents* that he is unable to provide an adequate discussion of 'the significance of work for the economics of the libido' (Freud 1953-74, 21.79). But his footnote does suggest what such a discussion might need to account for. On the one hand '[n]o other technique for the conduct of life attaches the individual so firmly to reality as laying emphasis on work': and yet work also offers the possibility 'of displacing a large amount of libidinal components, whether narcissistic, aggressive or even erotic', a possibility which 'lends it a value by no means second to what it enjoys as something indispensable to the preservation and justification of existence in society' (Freud 1953-74: 21.79). Work is in some indissociable way both actual and phantasmal – for it does its phantasmal work precisely in standing as the very epitome of the actual. If work is what machines do, then a 'work' also remains the word for what art is and does. But not the least of the perplexities attaching to the machinery of perpetuity is the perennial problem of what it means for it, and indeed for anyone or anything, to 'work'.

9

Shutdown

For some time we have been moving from an age of machines to an age of technology. The different work done by the two terms is instructive. If 'technology' is equivalent to 'machinery', there appears to be no singular term to fulfil the function of the word 'machine': 'a technology' does not refer to a specific piece of apparatus but generically to a whole family of machines, or a mechanical principle, like water-power, steam engines, or electronics. Machines nowadays rarely seem like entire or independent entities, but rather function as parts of networks. In order to be 'intelligent', machines need to become 'intelligencers', in the old sense of bringers of news or information. Intelligence signifies etymologically *inter* + *legere*, an act of reading or sorting between. 'Machine intelligence' is both a feature of certain mechanisms and what the users of intelligent machines must develop, or imagine, or develop through imagination – or perhaps it is just the exchange of intelligence between them. When the principal work done by any machine is to communicate with other machines, it becomes harder to conceive a machine as a free-standing image or object, and correspondingly more necessary to imagine its workings, through more or less satisfactory surrogates or synecdoches. If machines have always in some sense perturbed visibility, the steady transformation of objects into machines and the movement of all machines toward the condition of media may seem to exceed visibility altogether, or remove such machines from the realm of the visible to that of the visionary. If there are more devices and more apparatuses than ever, the work performed by technology seems more and more to have to be imaginary.

I have wanted to suggest that it is precisely this intimate association between the work of imagination and the workings of machines that makes imagining itself more and more apt to be conceived as a mechanical work, even if such activity must necessarily be the work-in-progress of a machine dreaming up its own workings. However, the move from objects to systems means that the problem, if it is one, is not so much that it is hard to imagine the workings of contemporary machines as that it is hard to imagine anything that could be thought to be in principle non-mechanical. From being a synonym for reduction and limit, mechanism has come to seem like the very form of the open. Machines have always had to be imagined, that is, made sense of, adjusted to, or put into operation, in the process I have called technesis, but the condition of a machine may now be essentially to-be-imagined. Machines used to be said to be 'in development', but perhaps all machines are now 'under imagination'. What Gilbert Simondon (2016) describes as the 'mode of existence' of machines has now become essentially prospective.

How does this matter? Technology will always seem to ask or provoke such a question and, no matter who articulates it, the question concerning technology will always seem to be a critical matter. To think about technology always seem to commit us to make a judgement upon or diagnosis of our technical condition, to discern some destiny, to curve analysis into prophecy. It nerves us in some way for change.

There is, by contrast, no such critical or polemical push in this book, which I have meant to be actuated as little as possible by any libido of ultimacy. I do not aim to show that we have arrived at or inhabit any particular kind of predicament, or to indicate the way to a better condition with relation to our phantasmal forms of technesis, or imaginary relation to technological forms. I intend neither to decry the workings of mechanical fantasy nor to affirm them as any intimation of utopia. I see neither catastrophe nor salvation in this history and its further unfolding. I have wanted merely to make our machine intelligence to some greater degree intelligible. We have always dreamed of and with machines. We are always dreaming through the machines we have been dreaming about, even if those kinds of machine and the kinds of dreamwork have been various and changeable. Such variations are not random fluctuations, or not entirely, and so there is certainly a kind of history to be made out of these dream-machines and machined dreams. Indeed, one may say that any understanding of the force and purport of technology must always bear on the question of time, since, as Bernard Stiegler has emphasised, all machines aim to control time and are themselves governed by the sense of a temporal horizon. All machines are therefore time machines, that operate in time and give it scansion, and history itself must pass through technology. But this temporality does not simply yield itself up, and is not simply available to be read off. The temporalities opened up by the technical are itself a part of the technesis, or dreamwork of the machine, not its simple effect or excrement.

The history of technology is not to be rounded up into an eschatology. It is what it is, even if what that is exactly will always in part remain to be seen. This is no indolent failure of nerve, no betrayal of the critical impulse, or spineless acceptance of the merely existing status quo. On the contrary, it might be said that our contemporary status quo is constituted by the easy surrender to the self-certifying glamour of critique. The libido of ultimacy might very well form part of our technical dreamwork, but that just means it is among the things about which we might wish more intelligence. Prophecy turns away from the object, turns away, precisely, from bothersome technicality.

We may well be at a turning point in technical matters, in relation in particular to energy sources and climate change. Where most other human societies have struggled against finitude, our struggle may be against our own institutionalised dreams of the limitless. Such questions are doubtless related to the potential dangers of our market structures and economic processes. But these are technical and not spiritual, cultural or even philosophical questions. This is to take a view opposite to that articulated by Heidegger in his 'The Question Concerning Technology' of 1954, in which he writes that 'the essence of technology is by no means anything technological [...] nothing on the order of a machine' (Heidegger 1977: 287, 305). By contrast, I would like you to believe that there is less than ever one necessary thing it means for something to be technical and so less than ever any kind of essence of the technological to be wrestled out, whether as the '[t]he rule of enframing [that] threatens man with the possibility that it could be denied to him to enter into a more original revealing and hence to experience the call of a more primal truth' (309), or as the 'saving power', in the phrase Heidegger borrows from Hölderlin (310), that will free us from the deadly danger of that enframing. We might similarly doubt that what Simondon calls 'the essence of technicity' is best understood in the way he proposes:

> The existence of technical objects and the conditions of their genesis pose for philosophical thought a question which cannot be resolved simply by a simple consideration of the objects themselves. What is the bearing of meaning of the genesis of technical objects on the totality of thought, the existence of man and his way of being in the world? (Simondon 1989: 154; my translation)

The absence of a prior and separable essence of technology is precisely the reason to think that we cannot get away from machines and the technical questions that arise from them. The ways in which we live our relations to machines must be seen as themselves technically formed and inflected, where we do not know in advance what 'technical' or 'technically' might have to mean. This is not a crippling predicament, since agnostic expertise and expert agnosticism is precisely what it must always mean for something to be technical. There is no way of framing ethical, political, and philosophical questions that would not also have to be a matter of techne, technique, or technesis, and so would not have to be imagined mechanically.

Works Cited

A.B. 1689. *Some Remarks Upon Government, and Particularly Upon the Establishment of the English Monarchy Relating to this Present Juncture in Two Letters.* London: n.p.

Abhedânanda, Swâmi. 1902. *How to Be a Yogi.* New York: Vedanta Society.

Adams, George. 1785. *An Essay on Electricity, Explaining the Theory and Practice of that Useful Science; and the Mode of Applying It to Medical Purposes.* 2nd edn. London: the Author.

Agrippa von Nettesheim, Heinrich Cornelius. 1651. *Three Books of Occult Philosophy*, translated by John French. London: for Gregory Moule.

'Alcoa: The Place They Do Imagineering'. 1942. *Time Magazine*, February 16: 59.

Alstead, Johann Heinrich. 1664. *Templum Musicum, or, The Musical Synopsis of the Learned and Famous Johannes-Henricus-Alstedius.* Translated by John Birchensha. London: Will. Godbid for Peter Dring.

'Ancient Tek' (Ancient Transformational Technologies). 2014. http://ancienttek.com/.

Anderson, Michael, and Susan Leigh Anderson, editors. 2011. *Machine Ethics.* Cambridge: Cambridge University Press.

Archibald, Timothy. 2005. *Sex Machines: Photographs and Interviews.* Los Angeles: Process.

Aristotle. 1995. *Poetics.* Edited and translated by Stephen Halliwell. Cambridge: Harvard University Press.

Arsić, Branka. 2005. 'Melville's Celibatory Machines: Bartleby, *Pierre* and "The Paradise of Bachelors"'. *Diacritics*, 35: 81-100.

Ascott, Roy. 2003. *Telematic Embrace: Visionary Theories of Art, Technology and Consciousness.* Berkeley: University of California Press.

Babbage, Charles. 1826. 'On a Method of Expressing by Signs the Action of Machinery'. *Philosophical Transactions*, 116: 250-65.

Babbage, Charles. 1864. *Passages from the Life of a Philosopher.* London: Longman, Green, Longman, Roberts and Green.

Ball, Geoff. n.d. 'Technography'. http://www.geoffballfacilitator.com/technography.html.

Ball, Philip. 2014. *Invisible: The Dangerous Allure of the Unseen*. London: Bodley Head.

Baraduc, Hippolyte. 1913. *The Human Soul: Its Movements, Its Lights, and the Iconography of the Fluidic Invisible*. Paris: G.A. Mann.

Barclay, Robert. 1679. *Robert Barclay's Apology for the True Christian Divinity Vindicated from John Brown's Examination and Pretended Confutation Thereof in His Book called Quakerisme the Pathway to Paganisme*. London: Benjamin Clerk.

Barrett, Deirdre. 2001. *The Committee of Sleep: How Artists, Scientists and Athletes Use Dreams for Creative Problem-Solving – and How You Can Too*. N.p.: Oneiroi Press.

Bataille, Georges. 2001. *Story of the Eye*. Translated by Joachim Neugroschal. London: Penguin.

Beer, Gillian. 1996. 'Authentic Tidings of Invisible Things: Vision and the Invisible in the Later Nineteenth Century'. In *Vision in Context: Historical and Contemporary Perspectives on Sight*, edited by Teresa Brennan and Martin Jay, 833-98. London: Routledge.

Bell, Vaughan, Ethan Grech, Cara Maiden, Peter W. Halligan and Hadyn D. Ellis, 2005. '"Internet Delusions": A Case Series and Theoretical Integration'. *Psychopathology*, 38: 144-50.

Benjamin, Walter. 2007. 'The Work of Art in the Age of Mechanical Reproduction'. In *Illuminations: Essays and Reflections*, translated by Harry Zohn, 217-52. New York: Schocken.

Bennett, Arnold. 1911. *The Human Machine*. New York: George Doran.

Bergson, Henri. 1911. *Laughter: An Essay on the Meaning of the Comic*. Translated by Cloudesley Brereton and Fred Rothwell. New York: Macmillan.

Bernoulli, Johannis. 1742. *Opera omnia*. 4 volumes. Lausanne: Marci-Michaelis Bousquet.

Bertucci, Paola. 2006. 'Revealing Sparks: John Wesley and the Religious Utility of Electrical Healing'. *British Journal for the History of Science*, 39: 341-62.

Bertucci, Paola. 2016. 'Shocking Subjects: Human Experiments and the Material Culture of Medical Electricity in Eighteenth-Century England'. In *The Uses of Humans in Experiment: Perspectives from the 17th to the 20th Century*, edited by Erika Dyck and Larry Stewart, 111-38. Leiden: Brill/Rodopi.

'Bicycle Erotica'. n.d. http://www.oldbike.eu/museum/1900s/bicycle-erotica/.

'Bike Sex Man Placed on Probation'. 2007. *BBC News*. http://news.bbc.co.uk/1/hi/scotland/glasgow_and_west/7095134.stm

Birch, John. 1803. *An Essay on the Medical Application of Electricity*. London: n.p.

Blackerby, Samuel. 1674. *Sermons Preached on Several Occasions*. London: for Nevil Simmons.

Blaustein, Richard. 1992. 'Traditional Healing Today: Moving Beyond Stereotypes'. In *Herbal and Magical Medicine: Traditional Healing Today*, edited by James Kirkland, Holly F. Mathews, C.W. Sullivan III, and Karen Baldwin, 32-40. Durham: Duke University Press.

Bloomfield Moore, Clara. 1893. *Keely and his Discoveries: Aerial Navigation*. London: Kegan Paul, Trench, Trübner and Co.

Bondio, Mariacarla Gadebusch. 2009. '*Daedalus* sive mechanicus: Automaten und Maschinen an der Schnittstelle zwischen Mechanik und Medizin'. *Sudhoffs Archive*, 93: 4-25.

Boswell, James. 2008. *The Life of Samuel Johnson*. Edited by David Womersley. London: Penguin.

Bovet, Richard. 1684. *Pandaemonium, or, The Devil's Cloyster Being a Further Blow to Modern Sadduceism, Proving the Existence of Witches and Spirits*. London: for J. Walthoe.

Bowie, Theodore R., editor. 1959. *The Sketchbook of Villard de Honnecourt*. Bloomington: Indiana University Press.

Brentano, Peter 1830. *Perpetual Motion: Explained as It Is Discovered in Nature*. London: William Verrinder.

Brewster, David. 1834. 'Of the Influence of Successive Impulses of Light Upon the Retina'. *London and Edinburgh Philosophical Magazine*, 3rd series, 4: 241-5.

Brockett, Oscar G., Margaret Mitchell, and Linda Harberger. 2010. *Making the Scene: A History of Stage Design and Technology in Europe and the United States*. San Antonio: Tobin Theatre Arts Fund.

Brody, Herb. 1992. 'The Pleasure Machine'. *Technology Review*, 95: 31-6.

Brown, John. 1678. *Quakerisme the Path-way to Paganisme, or, A View of the Quakers Religion*. Edinburgh: for John Cairns et. al.

Browne, Thomas. 1977. *The Major Works*. Edited by C.A. Patrides. Harmondsworth: Penguin.

Brückner, Burkhart. 2016. 'Animal Magnetism, Psychiatry and Subjective Experience in Nineteenth-Century Germany: Friedrich Krauß and his *Nothschrei*'. *Medical History*, 60: 19-36.

Bühlmann, Vera and Ludger Hovestadt, editors. 2014. *Domesticating Symbols: Metalithikum II*. Vienna: Ambra.

Caldwell, Roy C. 1993. '*Tristram Shandy*: Bachelor Machine'. *The Eighteenth Century*, 34: 103-114.

Calef, Robert. 1700. *More Wonders of the Invisible World, or, The Wonders of the Invisible World Display'd in Five Parts*. London: for Nath. Hillar and Joseph Collyer.

Caley, Abraham. 1683. *A Glimpse of Eternity Very Useful to Awaken Sinners, and to Comfort Saints*. London: for Thomas Parkhurst and G.B.

Campbell, Lily B. 1923. *Scenes and Machines on the English Stage During the Renaissance: A Classical Revival*. Cambridge: Cambridge University Press.

Campbell, Mary Baine. 2011. 'Speedy Messengers: Fiction, Cryptography, Space Travel, and Francis Godwin's *The Man in the Moone*'. *The Yearbook of English Studies*, 41: 190-204.

Campbell-Kelly, Martin. 2012. 'The ACE and the Shaping of British Computing'. In *Alan Turing's Electronic Brain: The Struggle to Build the ACE, the World's Fastest Computer*, edited by B. Jack Copeland, 149-72. Oxford: Oxford University Press.

Carlson, W. Bernard. 2013. *Tesla: Inventor of the Electrical Age*. Princeton: Princeton University Press.

Carrouges, Michel. 1976. *Les machines célibataires*. 2nd edn. Paris: Chêne.

Carver, Beci. 2016. 'Absolutist Slot Machines'. In *Writing, Medium, Machine: Modern Technographies*, edited by Sean Pryor and David Trotter, 178-190. London: Open Humanities Press.

Cecil, Paul, editor. 1996. *FLICKERS of the Dreamachine*. Hove: Codex.

Chabris, Christopher, and Daniel Simons. 2011. *The Invisible Gorilla and Other Ways our Intuition Deceives Us*. London: HarperCollins.

Chaucer, Geoffrey. 2008. *The Riverside Chaucer*, edited by F.N. Robinson. 3rd edn. Oxford: Oxford University Press.

Chauliac, Guy de. 1971. *The Cyrurgie of Guy de Chauliac*, edited by Margaret S. Ogden. London: Oxford University Press for the Early English Text Society.

Clarke, Bruce. 2002. 'Mediating *The Fly*: Posthuman Metamorphosis in the 1950s'. *Configurations*, 10: 169-91.

Coffin, Judith G. 1994. 'Credit, Consumption and Images of Women's Desires: Selling the Sewing Machine in Late Nineteenth-Century France'. *French Historical Studies*, 18: 749-83.

Collier, Jeremy. 1699. *A Defence of the Short View of the Profaneness and Immorality of the English Stage* London: for S. Keble, R. Sare and H. Hindmarsh.

Compos, Luis A. 2015. *Radium and the Secret of Life*. Chicago: University of Chicago Press.

Connor, Steven. 2000. *Dumbstruck: A Cultural History of Ventriloquism*. Oxford: Oxford University Press.

Connor, Steven. 2008. 'Elan Mortel: Life, Death and Laughter'. http://stevenconnor.com/elanmortel.html.

Connor, Steven. 2009. 'Absolute Levity'. *Comparative Critical Studies*, 6: 411-27.

Connor, Steven. 2010. *The Matter of Air: Science and Art of the Ethereal*. London: Reaktion.

Connor, Steven. 2012. 'Soft Machines: Looking Through X-rays'. In *The Moderns: Wie sich das 20. Jahrhundert in Kunst und Wissenschaft erfunden hat*, edited by Cathrin Pichler and Susanne Neuburger, 207-18. Vienna: Springer.

Connor, Steven. 2015. 'Guys and Dolls'. *Women: A Cultural Review*, 26: 129-42.

Connor, Steven. 2016. 'How to do Things with Writing Machines'. In *Writing, Medium, Machine: Modern Technographies*, edited by Sean Pryor and David Trotter, 18-34. London: Open Humanities Press.

Connor, Steven. 2016. *Living by Numbers: In Defence of Quantity*. London: Reaktion.

Conrad, Joseph. 1983. *Collected Letters. Vol. 1: 1861-1897*. Edited by Frederick R. Karl and Laurence Davies. Cambridge: Cambridge University Press.

Copeland, B. Jack, editor. 2012a. *Alan Turing's Electronic Brain: The Struggle to Build the ACE, the World's Fastest Computer*. Oxford: Oxford University Press.

Copeland, B. Jack. 2012b. 'The Origins and Development of the ACE Project'. In *Alan Turing's Electronic Brain: The Struggle to Build the ACE, the World's Fastest Computer*, edited by B. Jack Copeland, 173-92. Oxford: Oxford University Press.

Corelli, Marie. 1886. *A Romance of Two Worlds*. 2 volumes. London: Richard Bentley.

Corelli, Marie. 1910. *The Devil's Motor: A Fantasy*. London: Hodder and Stoughton.

Corelli, Marie. 1911. *The Life Everlasting: A Romance of Reality*. New York: A.L. Burt.

Crooke, Helkiah. 1615. *Mikrokosmographia a Description of the Body of Man*. London: William Iaggard.

Cushing, Ellen. 2013. 'Amazon Mechanical Turk: The Digital Sweatshop'. *Utne* (January/February). http://www.utne.com/science-and-technology/amazon-mechanical-turk-zm0z13jfzlin.aspx.

Cutliffe, Stephen H. 2007. 'The Greenwood *Technographies*: Life Stories of Technologies'. *Technology and Culture*, 48: 165-8.

Daly, Herman E. 1986. 'The Economic Thought of Frederick Soddy'. In *Frederick Soddy (1877–1956): Early Pioneer in Radiochemistry*, edited by George B. Kaufman, 199-219. Dordrecht: D. Reidel.

Davidson, Clifford. 1996. *Technology, Guilds, and Early English Drama*. Kalamazoo: Medieval Institute Publications.

De Certeau, Michel. 1984. *The Practice of Everyday Life*. Translated by Steven Rendall. Berkeley: University of California Press.

Deleuze, Gilles, and Félix Guattari. 2000. *Anti-Oedipus: Capitalism and Schizophrenia*. Translated by Robert Hurley, Mark Seem, and Helen R. Lane. Minneapolis: University of Minnesota Press.

DeMeo, James. 1989. *The Orgone Accumulator Handbook: Construction Plans, Experimental Use and Protection Against Toxic Energy*. El Cerrito: Natural Energy Works.

Denning, Peter J., editor. 2002. *The Invisible Future: The Seamless Integration of Technology into Everyday Life*. New York: McGraw-Hill.

Dircks, Henry. 1870. *Perpetuum Mobile: or, A History of the Search for Self-Motive Power, from the 13th to the 19th Century*. 2nd edn. London: E. and F.N. Spon.

Douglas-Fairhurst, Robert. 2015. *The Story of Alice: Lewis Carroll and The Secret History of Wonderland*. Cambridge: Harvard University Press.

Duchamp, Marcel (1969). *Notes and Projects for The Large Glass*. Edited by Arturo Schwartz. Translated by George H. Hamilton, Cleve Gray, and Arturo Schwartz. London: Thames and Hudson.

Duchamp, Marcel. 1973. *Salt Seller: The Writings of Marcel Duchamp (Marchand du Sel)*. Edited by Michel Sanouillet and Elmer Peterson. Translated by Elmer Peterson et. al. New York: Oxford University Press.

Duffy, W.L. 2010. 'Monomania and Perpetual Motion: Insanity and Amateur Scientific Enthusiasm in Nineteenth-Century Medical, Scientific and Literary Discourse'. *French Cultural Studies*, 21: 155-66.

Ellis, Havelock. 1910. *Studies in the Psychology of Sex: Vol. 1 The Evolution of Modesty, The Phenomena of Sexual Periodicity, Auto-Erotism*. 3rd edn. Philadelphia. F.A. Davis Co.

Erdman, David V. 1977. *Blake: Prophet against Empire*. 3rd edn. Princeton: Princeton University Press.

Erskine, Ebenezer. 1791. *The Whole Works of the Rev. Mr. Ebenezer Erskine*. 2 volumes. Falkirk: Hugh Mitchell.

Espinoza, Tania. 2013. 'The Technical Object of Psychoanalysis'. In *Stiegler and Technics*, edited by Christina Howell and Gerald Moore, 151-64. Edinburgh: Edinburgh University Press

Evans, Cadwallader. 1776. 'A Relation of a Cure Performed by Electricity'. *Medical Observations and Inquiries*, 2nd edn. 1: 83-6.

Fessenden, Thomas Green. 1803. *Terrible Tractoration!! A Poetical Petition Against Galvanising Trumpery, and the Perkinistic Institution*. London: T. Hurst.

Fountain, John. 1661. *The Rewards of Vertue: A Comedie*. London: for Henry Fletcher.

Freud, Sigmund. 1953-74. *The Standard Edition of the Complete Psychological Works of Sigmund Freud*. 24 volumes. Edited and translated by James Strachey et. al. London: Hogarth Press.

Freud, Sigmund. 1991. *Gesammelte Werke*. 18 Vols. London: Imago.

Freud, Sigmund and Joseph Breuer. 1991. *Studies on Hysteria*. Translated by James and Alix Strachey. Harmondsworth: Penguin.

Fuchs, Thomas. 2006. 'Being a Psycho-Machine: On the Phenomenology of the Influencing Machine'. In *Air Loom: Der Luft-Webstuhl und andere gefährliche Beeinflussungsapparate/ The Air Loom and Other Dangerous Influencing Machines*, 25-41. Heidelberg: Verlag Das Wunderhorn.

Galasso, Joseph A. 1963. 'What's in a Name'. *STWP Review*, 10: 23-4.

Geley, Gustav. 1927. *Clairvoyance and Materialisation: A Record of Experiments*. Translated by Stanley De Brath. London: T. Fisher Unwin.

Gilman, Sander. 2008. 'Electrotherapy and Mental Illness: Then and Now'. *History of Psychiatry*, 19: 339-57.

'Glacier Milk'. 2014. http://ancienttek.com/GlacierMilk/Glacier_milk.htm.

Glasser, Otto. 1933. *Wilhelm Conrad Röntgen and the Early History of the Roentgen [sic] Rays: With A Chapter 'Personal Reminiscences of W.C. Röntgen' By Margret Boveri*. London: John Bale, Sons and Danielsson.

Godwin, Francis. 1638. *The Man in the Moone: Or, A Discourse of a Voyage Thither. By Domingo Gonsales. The Speedy Messenger*. London: John Norton.

Goff, Alan. 2006. 'Quantum Tic-Tac-Toe: A Teaching Metaphor for Superposition in Quantum Mechanics'. *American Journal of Physics*, 74: 962-73.

Gower, John. 1901. *Complete Works*, 4 volumes. Edited by G.C. Macaulay. Oxford: Clarendon Press.

Graham, James. 1778. *The General State of Medical and Chirurgical Practice Exhibited; Showing Them to Be Inadequate, Ineffectual, Absurd, and Ridiculous*. Bath: n.p.

Graham, James. 1782. *Il Convito Amoroso! Or, A Serio-comico-philosophical Lecture on the Causes, Nature, and Effects of Love and Beauty, at the Different Periods of Human Life, in Persons and Personages, Male, Female, and Demi-Charactère ; and in Praise of the Genial and Prolific Influences of the Celestial Bed*. 2nd edn. London: for Hebe Vestina.

Gramantieri, Riccardo. 2016. 'Re-emergence of the Death Instinct in Wilhelm Reich's Last Experiment'. *Psychoanalysis and History*, 18: 203-20.

Guerard, Albert J. 1969. 'The Vanishing Anarchists: A Memory'. *Sewanee Review*, 77: 440-62.

Guillemeau, Jacques. 1598. *The Frenche chirurgerye, or all the manualle operations of chirurgerye*. Translated by A.M. Dort: Isaac Canin

Gunning, Tom. 2012. ' "We are Here and Not Here": Late Nineteenth-Century Stage Magic and the Roots of Cinema in the Appearance (and Disappearance) of the Virtual Image'. In *A Companion to Early Cinema*, edited by André Gaudreault, Nicolas Dulac, and Santiago Hidalgo, 52-63. Oxford: John Wiley.

Gurney, Edmund, Frederic W.H. Myers, and Frank Podmore. 1886. *Phantasms of the Living*. 2 volumes. London: Society for Psychical Research/Trübner and Co.

Habermas, Jürgen. 1984. *The Theory of Communicative Action: Volume 1. Reason and the Rationalization of Society*. Translated by Thomas McCarthy. Cambridge: Polity Press.

Hamilton, Mary. 1906. *Incubation: Or, The Cure of Disease in Pagan Temples and Christian Churches*. St Andrews: W.C. Henderson and Son.

Hansen, Mark. 2000. *Embodying Technesis: Technology Beyond Writing*. Ann Arbor: University of Michigan Press.

Harrington, Walter. 1924. 'Making Clothes by Machine'. In *A Popular History of American Invention*, 2 volumes. Edited by Waldemar Kaempffert, 375-403. New York: Charles Scribner's Sons.

Harris, Paul A. 1997. 'Exploring Technographies: Chaos Diagrams and Oulipian Writing as Virtual Signs'. In *Reading Matters: Narrative in the New Media Ecology*, edited by Joseph Tabbi and Michael Wutz, 136-53. Ithaca: Cornell University Press.

Hearne, Keith. 1981. 'A "Light-Switch Phenomenon" in Lucid Dreams'. *Journal of Mental Imagery*, 5: 97-100.

Hearne, Keith. 1990. *The Dream Machine: Lucid Dreams and How to Control Them*. Wellingborough: Aquarian Press.

Heath, Malcolm. 2004. 'Technography'. In *Menander: A Rhetor in Context*, 255-76. Oxford: Oxford University Press.

Heidegger, Martin. 1977. 'The Question Concerning Technology'. Translated by William Lovitt. In *Martin Heidegger: Basic Writings*, edited by David Farrell Krell, 287-317. New York: Harper and Row.

Heilbron, J.L. 1979. *Electricity in the 17th and 18th Centuries: A Study of Early Modern Physics*. Berkeley: University of California Press.

Henderson, Lynda Dalrymple. 1998. *Duchamp in Context: Science and Technology in the Large Glass and Related Works*. Princeton: Princeton University Press.

Hering, Philip W. 1924. *Foibles and Fallacies of Science: An Account of Celebrated Scientific Vagaries.* London: George Routledge and Sons.

Hervey de Saint-Denys, Marie Jean Léon, Marquis d'. 1867. *Les Rêves et les moyens de les diriger: Observations pratiques.* Paris: Amyot.

Hirjak, Dusan, and Thomas Fuchs. 2010. 'Delusions of Technical Alien Control: A Phenomenological Description of Three Cases'. *Psychopathology*, 43: 96-103.

Hooke, Robert. 1665. *Micrographia: or, Some Physiological Descriptions of Minute Bodies Made by Magnifying Glasses.* London: John Martyr and James Allestry.

Hughes, Ted. 2003. *Collected Poems.* Edited by Paul Keegan. London: Faber and Faber.

Humboldt, Wilhelm von. 1999. *On Language: On the Diversity of Human Language Construction and its Influence on the Mental Development of the Human Species.* 2nd edn. Edited by Michael Losonsky. Translated by Peter Heath. Cambridge: Cambridge University Press.

'Imagineers, The'. 1996. *Walt Disney Imagineering: A Behind the Dreams Look at Making the Magic Real.* New York: Hyperion.

Jádi, Ferenc. 2006. 'Acquainting Oneself with the Truth of Allegory: Heintzen's Self-Devised Interpretation of his Character Transformation'. In *Air Loom: Der Luft-Webstuhl und andere gefährliche Beeinflussungsapparate/The Air Loom and Other Dangerous Influencing Machines*, 197-213. Heidelberg: Verlag Das Wunderhorn.

Jansen, Kees and Sietze Vellema. 2011. 'What Is Technography?' *NJAS - Wageningen Journal of Life Sciences*, 57: 169–177.

Jay, Mike. 2003. *The Air Loom Gang: The Strange and True Story of James Tilly Matthews and His Visionary Madness.* London: Bantam.

Jollie, Thomas. 1698. *A Vindication of the Surey Demoniack as No Impostor.* London: for Nevill Simmons.

Kafka, Franz. 1993. *Collected Stories.* Translated by Willa and Edwin Muir. New York: Alfred A. Knopf.

Kahn, Douglas. 2004. 'A Musical Technography of John Bischoff'. *Leonardo Music Journal*, 14: 74-79.

Kant, Immanuel. 2012. *Groundwork of the Metaphysic of Morals.* Edited and translated by Mary Gregor and Jens Timmermann. Cambridge: Cambridge University Press.

Keller, Evelyn Fox. 2007. 'Whole Bodies, Whole Persons? Cultural Studies, Psychoanalysis, and Biology'. In *Subjectivity: Ethnographic Investigations*, edited by João Biehl, Byron Good, and Arthur Kleinman, 352-61. Berkeley: University of California Press.

Kelly, Donald A. 1991. *The Manual of Free Energy Devices and Systems*. Clayton Cadake Industries and TriState Press. http://www.free-energy-info.co.uk/DonKelly.pdf.

Kerzner, Harold. 2014. 'Disney: Imagineering Project Management'. http://www.drharoldkerzner.com/disney-imagineering-project-management/.

Kien, Grant. 2008. 'Technography = Technology + Ethnography'. *Qualitative Inquiry*, 14: 1101-09.

King, Serge. 1981. *Imagineering for Health: Self-Healing Through the Use of the Mind*. Wheaton: Theosophical Publishing House.

Knoespel, Kenneth J. 1992. 'Gazing on Technology: *Theatrum Mechanorum* and the Assimilation of Renaissance Machinery'. In *Literature and Technology*, edited by Mark L. Greenberg and Lance Schachterle, 99-124. Bethlehem: Lehigh University Press.

Kraus, Alfred. 1994. 'Phenomenology of the Technical Delusion in Schizophrenics'. *Journal of Phenomenological Psychology*, 25: 51-69.

Kraus, Manfred. 2011. 'How to Classify Means of Persuasion: The *Rhetoric to Alexander* and Aristotle on *Pisteis*'. *Rhetorica: A Journal of the History of Rhetoric*, 29: 263-79.

Krauß, Friedrich. 1852. *Nothschrei eines Magnetisch-Vergifteten; Thatbestand, erklärt durch ungeschminkte Beschreibung des 36jährigen Hergangs, belegt mit allen Beweisen und Zeugnissen. Zur Belehrung und Warnung besonders für Familienväter und Geschäftsleute*. Stuttgart: the author.

Krauß, Friedrich. 1867. *Nothgedrungene Fortsetzung meines Nothschrei gegen meine Vergiftung mit concentrirtem Lebensäther und gründliche Erklärung der maskirten Einwirkungsweise desselben auf Geist und Körper zum Scheinleben*. Stuttgart: the author.

Krauß, Friedrich. 1967. *Nothschrei eines Magnetisch-Vergifteten: Selbstschilderungen eines Geisteskranken*. Edited by Heinz Ahlenstiel and Joachim Ernst Meyer. Leverkusen: Bayer.

Krieg, Eric. 2002. *Eric's History of Perpetual Motion and Free Energy Machines*. http://u2.lege.net/newebmasters.com__freeenergy/external_links_from_phact.org/dennis4.html.

LaBelle, Brandon. 2014. *Lexicon of the Mouth: Poetics and Politics of Voice and the Oral Imaginary*. New York: Bloomsbury.

Langelaan, George. 1959. *The Masks of War*. Garden City: Doubleday.

Langelaan, George. 1964. *Out of Time*. London: New English Library.

Latour, Bruno. 1993. *We Have Never Been Modern*. Translated by Catherine Porter. Cambridge: Harvard University Press.

Latour, Bruno. 1997. 'Trains of Thought: Piaget, Formalism, and the Fifth Dimension'. *Common Knowledge* 6: 170-91.

Lennon, John. 2015. *A Spaniard in the Works*. Edinburgh: Canongate.

'L'érotic Bicycle Film Festival débarque à Paris le 10 Septembre 2011'. 2011. *Carfee.fr* http://carfree.free.fr/index.php/2011/08/31/lerotic-bicycle-film-festival-debarque-a-paris-le-10-septembre-2011/.

Levins, Hoag. 1996. *American Sex Machines: The Hidden History of Sex at the US Patent Office*. Holbrook: Adams Media Corporation.

Lewin, Bertram D. 1946. 'Sleep, the Mouth, and the Dream Screen'. *Psychoanalytic Quarterly*, 15: 419-34.

Lippit, Akira Mizuta. 2005. *Atomic Light (Shadow Optics)*. Minneapolis: University of Minnesota Press.

Loeb, Lori. 1999. 'Consumerism and Commercial Electrotherapy: The Medical Battery Company in Nineteenth-Century London'. *Journal of Victorian Culture*, 4: 252-75.

Lovett, Richard. 1756. *The Subtil Medium Prov'd: or, That Wonderful Power of Nature, so Long Ago Conjectur'd by the Most Ancient and Remarkable Philosophers, Which They Call'd Sometimes Æther, but Oftener Elementary Fire, Verify'd*. London: for J. Hinton, W. Sandby and R. Lovett.

Lovett, Richard. 1760. *The Reviewers Review'd; or, The Bush-fighters Exploded: Being a Reply to the Animadversions, Made by the Authors of the Monthly Review, on a Late Pamphlet, Entitled Sir Isaac Newton's Æther Realiz'd*. Worcester: the Author.

MacLeish, Archibald. 1985. *New and Collected Poems 1917-1982*. Boston: Houghton Mifflin.

Maines, Rachel P. 1999. *The Technology of Orgasm: "Hysteria," the Vibrator, and Women's Sexual Satisfaction*. Baltimore: Johns Hopkins University Press.

Marcuse, Herbert. 2002. *One Dimensional Man: Studies in the Ideology of Advanced Industrial Society*. Abingdon: Routledge.

Marcuse, Herbert. 2005. *Eros and Civilization: A Philosophical Inquiry into Freud*. Abingdon: Routledge.

Martin, William. 1821. *A New System of Natural Philosophy, on the Principle of Perpetual Motion; with a Variety of Other Useful Discoveries*. Newcastle: Preston and Heaton.

McLuhan, Marshall. 1994. *Understanding Media: The Extensions of Man*. Cambridge: MIT Press.

Medulla Grammatice. n.d. Stonyhurst College ms. XV A.1.10.

Melton, J. Gordon, editor. 2001. *Encyclopedia of Occultism and Parapsychology*. 5th edn. 2 volumes. Detroit: Gale.

Milton, John. 1998. *Complete Poems*. Edited by John Leonard. London: Penguin.

Mindell, David A. 1995. 'Automation's Finest Hour: Bell Labs and Automatic Control in World War II'. *IEEE Control Systems*, 15: 72-80.

Mollmann, Stephen. 2010. '*The War of the Worlds* in the *Boston Post* and the Rise of American Imperialism: "Let Mars Fire"'. *English Literature in Transition 1880-1920*, 53: 387-412.

More, Henry. 1642. ΨΥΧΩΔΙΑ *Platonica, or, A Platonicall Song of the Soul*. Cambridge: Roger Daniel.

Morus Iwan Rhys. 1998. *Frankenstein's Children: Electricity, Exhibition, and Experiment in Early-Nineteenth-Century London*. Princeton: Princeton University Press.

Morus, Iwan Rhys. 1999. 'The Measure of Man: Technologizing the Victorian Body'. *History of Science*, 37: 249-82.

Morus, Iwan Rhys. 2011. *Shocking Bodies: Life, Death and Electricity in Victorian England*. Stroud: History Press.

Moshenska, Joe. 2014. *Feeling Pleasures: The Sense of Touch in Renaissance England*. Oxford: Oxford University Press.

Mousoutzanis, Aris. 2014. *Fin-de-Siècle Fictions, 1890s-1990s: Apocalypse, Technoscience, Empire*. Houndmills: Palgrave Macmillan.

Nelson, Theodor H. 2003. 'From *Computer Lib/Dream Machines*'. In *The New Media Reader*, edited by Noah Wardrip-Fruin and Nick Montfort, 301-38. Cambridge: MIT Press.

Newton, Isaac. 1704. *Opticks: or, a Treatise of the Reflexions, Refractions, Inflexions and Colours of Light*. London: for Samuel Smith and Benjamin Walford.

Novalis (Friedrich von Hardenberg). 1997. 'Christendom or Europe?' In *Philosophical Writings*, translated and edited by Margaret Mahony Stoljar, 137-52. Albany: State University of New York Press.

Nozick, Robert. 2003. *Anarchy, State, and Utopia*. Malden: Blackwell.

Numerico, Teresa. 2012. 'From Turing Machine to "Electronic Brain"'. In *Alan Turing's Electronic Brain: The Struggle to Build the ACE, the World's Fastest Computer*, edited by B. Jack Copeland, 33-92. Oxford: Oxford University Press.

O'Brien, Flann. 1974. *The Third Policeman*. London: Picador.

Offen, Karen. 1998. ' "Powered by a woman's foot:" A Documentary Introduction to the Sexual Politics of the Sewing Machine in Nineteenth-Century France'. *Women's Studies International Forum*, 11: 93-101.

Olshin, Benjamin B. 2009. 'Sophistical Devices: Leonardo da Vinci's Investigations of Perpetual Motion'. *Icon*, 15: 1-39.

Paijmans, Theo. 2004. *Free Energy Pioneer: John Worrell Keely*. Kempton: Adventures Unlimited Press.

Park, David. 1997. *The Fire within the Eye: A Historical Essay on the Nature and Meaning of Light*. Princeton: Princeton University Press.

Perceval, John. 1962, *Perceval's Narrative: A Patient's Account of his Psychosis 1830-1832*. Edited by Gregory Bateson. London: Hogarth Press

Peña, Carolyn Thomas de la. 2003. *The Body Electric: How Strange Machines Built the Modern American*. New York: New York University Press.

Peyrière, Monique. 2007. 'Femmes au travail, machines en chaleur: l'emprise de la machine à coudre'. *Communications*, 81: 71-84.

Plato. 1998. *The Republic*. Translated by Robin Waterfield. Oxford: Oxford University Press.

Poe, Edgar Allan. 1836. 'Maelzel's Chess-Player'. *Southern Literary Messenger*, 2: 318-326.

Poulliet, Thésée. 1887. *Etude médico-philosophique sur les forms, les causes, les signes, les conséquences et le traitement de l'onanisme chez la femme*. 5th edn. Paris: A. Delahaye and E. Lecrosnier.

Pound, Ezra. 1970. *Guide to Kulchur*. New York: New Directions.

Pryor, Sean and David Trotter. 2016. 'Introduction'. In *Writing, Medium, Machine: Modern Technographies*, edited by Sean Pryor and David Trotter, 7-17. London: Open Humanities Press.

'Public Amusements'. 1865. *Lloyd's Weekly Newspaper*, June 18: 8.

Punt, Michael. 2000. *Early Cinema and the Technological Imaginary*. Amsterdam: Postdigital Press.

Purdon, James. 2016. 'Texts, Technics, Technographies'. Unpublished lecture, given at the TextTechniques workshop, Universität Erfurt, January 13.

Pynchon, Thomas. 1982. *The Crying of Lot 49*. Toronto, New York: Bantam.

Radcliffe, Anne. 1986. *The Romance of the Forest*. Edited by Chloe Chard. Oxford: Oxford University Press.

Read, John. 1995. *From Alchemy to Chemistry*. New York: Dover.

Reich, Wilhelm. 1956. 'Re-emergence of Freud's "Death Instinct" as "DOR" Energy'. *Orgonomic Medicine*, 2: 2-11.

Reich, Wilhelm. 1946. *The Mass Psychology of Fascism*. 3rd edn. Translated by Theodore P. Wolfe. New York: Orgone Institute Press.

Reich, Wilhelm. 1974. *The Function of the Orgasm: Sex-Economic Problems of Biological Energy* Translated by Vincent R. Carfagno. New York: Farrar, Straus and Giroux.

Reich, Wilhelm. 1999. *American Odyssey: Letters and Journals 1940-1947*. Edited by Mary Boyd Higgins. New York: Farrar, Straus and Giroux.

Reich, Wilhelm, Roger du Teil, and Arthur Hahn (1938). *Die Bione: zur Entstehung des vegetativen Lebens*. Oslo: Sexpol-Verlag.

Reichenbach, Carl von. 1926. *Reichenbach's Letters on Od and Magnetism (1852). Published for the First Time in English, With Extracts from His Other Works, So as to Make a Complete Presentation of the Odic Theory*. Translated by F.D. O'Byrne. London: Hutchinson and Co.

Reuleaux, Franz. 1876. *The Kinematics of Machinery: Outlines of a Theory of Machines*. Translated by Alex B.W. Kennedy. London: Macmillan and Co.

Richards, I.A. 1965. *The Philosophy of Rhetoric*. Oxford: Oxford University Press.

Richet, Charles. 1923. *Thirty Years of Psychical Research: Being a Treatise on Metaphysics*. Translated by Stanley De Brath. New York: Macmillan.

Riskin, Jessica. 2003. 'The Defecating Duck, or, the Ambiguous Origins of Artificial Life'. *Critical Inquiry*, 29: 599-633.

Röske, Thomas, and Bettina Brand-Claussen. 2006. 'Illustrations of Madness: Delusions, Machines and Art'. In *Air Loom: Der Luft-Webstuhl und andere gefährliche Beeinflussungsapparate/ The Air Loom and Other Dangerous Influencing Machines*, 9-23. Heidelberg: Verlag Das Wunderhorn.

Roussel, Raymond. 1983. *Locus Solus*. Translated by Rupert Copeland Cuningham. London: John Calder.

Sartre, Jean-Paul. 1984. *Being and Nothingness: An Essay on Phenomenological Ontology*. Translated by Hazel E. Barnes. London: Methuen.

Sass, Louis A. 1995. *The Paradoxes of Delusion: Wittgenstein, Schreber and the Schizophrenic Mind*. Ithaca: Cornell University Press.

Sawday, Jonathan. 2007. *Engines of the Imagination: Renaissance Culture and the Rise of the Machine*. Abingdon: Routledge.

Schaffer, Simon. 1995. 'The Show That Never Ends: Perpetual Motion in the Early Eighteenth Century'. *British Journal for the History of Science*, 28: 157-89.

Scheerbart, Paul. 2011. *The Perpetual Motion Machine: The Story of an Invention*. Translated by Andrew Joron. Cambridge: Wakefield Press.

Schreber, Daniel Paul. 2000. *Memoirs of My Nervous Illness*. Translated and edited by Ida Macalpine and Richard A. Hunter. New York: New York Review Books.

Schrenk-Notzing, Baron von. 1923. *Phenomena of Materialisation: A Contribution to the Investigation of Mediumistic Teleplastics*. Translated by E.E. Fournier d'Albe. London: Kegan Paul, Trench, Trubner and Co.

Sconce, Jeffrey. 2011. 'On the Origins of the Origins of the Influencing Machine'. In *Media Archaeology: Approaches, Applications, and Implications*, edited by Erkki Huhtamo and Jussi Parikka, 70-94. Berkeley: University of California Press.

Serres, Michel. 1982a. 'The Origin of Language: Biology, Information Theory, and Thermodynamics'. In *Hermes: Literature, Science, Philosophy*, edited by Josué A. Harari and David F. Bell, various translators, 71-83. Baltimore: Johns Hopkins University Press.

Serres, Michel. 1982b. 'Turner Translates Carnot'. In *Hermes: Literature, Science, Philosophy*, edited by Josué A. Harari and David F. Bell, various translators, 54-62. Baltimore: Johns Hopkins University Press.

Serres, Michel. 2008. '*Feux et Signaux de Brume*: Virginia Woolf's Lighthouse'. Translated by Judith Adler. *SubStance*, 37: 110-31.

Serres, Michel. 2014. *Pantopie: de Hermès à Petite Poucette: Entretiens avec Martin Legros et Sven Ortoli*. Paris: Le Pommier.

Serviss, Garrett P. 1947. *Edison's Conquest of Mars*. Los Angeles: Carcosa House.

Sextus Empiricus. 1998. *Against the Grammarians*. Translated by D.L. Blank. Oxford: Clarendon Press.

Shakespeare, William. 2005. *Hamlet*. Edited by Ann Thompson and Neil Taylor. London: Bloomsbury.

Shakespeare, William. 2007. *Shakespeare's Poems: Venus and Adonis, The Rape of Lucrece and The Shorter Poems*. Edited by Katherine Duncan-Jones and Henry Woudhuysen. London: Arden Shakespeare.

Shakespeare, William. 2015. *Macbeth*. Edited by Sandra Clark and Pamela Mason. London: Bloomsbury Arden.

Sharp, Lesley A. 2014. *The Transplant Imaginary: Mechanical Hearts, Animal Parts, and Moral Thinking in Highly Experimental Science*. Berkeley: University of California Press.

Sheehan, Nik, dir. 2007. *FLicKeR*. National Film Board of Canada. https://www.youtube.com/watch?v=1JFgNMVePaQ.

Shelley, Percy Bysshe. 2004. *Complete Poetry. Vol. 2*. Edited by Donald H. Reiman, Neil Fraistat, and Nora Crook. Baltimore: Johns Hopkins University Press.

Sidney, Philip. 2002. *An Apology for Poetry (or the Defence of Poesy)*. 3rd edn. Edited by Geoffrey Shepherd and R.W. Maslen. Manchester: Manchester University Press.

Siegert, Bernhard. 2015. *Cultural Techniques: Grids, Filters, Doors, and Other Articulations of the Real*. Translated by Geoffrey Winthrop-Young. Stanford: Stanford University Press.

Simondon, Gilbert. 1989. *Du mode d'existence des objets techniques*. Paris: Aubier.

Sloterdijk, Peter. 2004. *Schäume: Sphären, Vol. 3: Plurale Sphärologie*. Frankfurt: Suhrkamp.

Smith, Kirby Flower. 1902. 'The Tale of Gyges and the King of Lydia'. *American Journal of Philology*, 23: 261-82, 361-87

Soddy, Frederick. 1909. *The Interpretation of Radium*. London: J. Murray.

Soddy, Frederick. 1922. *Cartesian Economics: The Bearing of Physical Science Upon State Stewardship*. London: Henderson.

Spengler, Oswald. 1926. *The Decline of the West: Form and Actuality*. 2 volumes. Translated by Charles Francis Atkinson. New York: Alfred A. Knopf.

Spengler, Oswald. 1932. *Man and Technics: A Contribution to a Philosophy of Life*. Translated by Charles Francis Atkinson. New York: Alfred A. Knopf.

Stead, W.T. 1893. 'Throughth; Or, On the Eve of the Fourth Dimension: A Record of Experiments in Telepathic Automatic Handwriting'. *Review of Reviews*, 7: 426-32

Steigerwald, Joan. 2016. 'The Subject as Instrument: Galvanic Experiments, Organic Apparatus and Problems of Calibration'. In *The Uses of Humans in Experiment: Perspectives from the 17th to the 20th Century*, ed. Erika Dyck and Larry Stewart, 80-110. Leiden: Brill/Rodopi.

Steinmeyer, Jim. 2005. *Hiding The Elephant: How Magicians Invented the Impossible*. London: Arrow Books.

Stiegler, Bernard. 1998. *Technics and Time, 1: The Fault of Epimethus*. Translated by Richard Beardsworth and George Collins. Stanford: Stanford University Press.

Stoppard, Tom, and Clive Exton. 1991. *The Boundary*. London: Samuel French.

Strathern, Paul. 2001. *Mendeleyev's Dream: The Quest for the Elements*. New York: Thomas Dunne.

Sullivan, Walter. 1962. 'The South Pole Fifty Years After'. *Arctic*, 15: 175-8.

Syson, Lydia. 2008. *Doctor of Love: James Graham and his Celestial Bed*. London: Alma Books.

Tausk, Victor. 1933. 'On the Origin of the "Influencing Machine" in Schizophrenia'. Translated by Dorian Feigenbaum. *Psychoanalytic Quarterly*, 2: 519-56.

ter Meulen, B.C., D. Tavy, and B.C. Jacobs. 2009. 'From Stroboscope to Dream Machine: A History of Flicker-Induced Hallucinations'. *European Neurology*, 62: 316-20.

Tesla, Nikola. 1900. 'The Problem of Increasing Human Energy'. *Century Illustrated Magazine*, 60: 175-210.

The Doctrine of Devils Proved to be the Grand Apostacy of These Later Times. An Essay Tending to Rectifie Those Undue Notions and Apprehensions Men Have About Daemons and Evil Spirits. 1676. London: for the Author.

'The Tagarene Shop'. 1956. *Shipbuilding and Shipping Record*, 87: 422.

Thompson, Sylvanus P. 1888. 'The Influence Machine, From 1788 to 1888'. *Journal of the Society of Telegraph-Engineers and Electricians*, 17: 569-628.

Thurschwell, Pamela. 2001. *Literature, Technology and Magical Thinking, 1880-1920*. Cambridge: Cambridge University Press.

Tibbits, Herbert. 1879. *How to Use a Galvanic Battery in Medicine and Surgery*. 2nd edn. London: J. and A. Churchill.

Trotter, David. 2001. *Paranoid Modernism: Literary Experiment, Psychosis, and the Professionalization of English Society*. Oxford: Oxford University Press.

Trotter, David. 2013. *Literature in the First Media Age: Britain Between the Wars*. Cambridge: Harvard University Press.

Trower, Shelley. 2012. *Senses of Vibration: A History of the Pleasure and Pain of Sound*. London: Continuum.

Turing, A.M. 1945. 'Proposed Electronic Calculator'. http://www.alanturing.net/turing_archive/archive/p/p01/p01.php.

Turner, Chris. 2011. *Adventures in the Orgasmatron: Wilhelm Reich and the Invention of Sex*. London: Fourth Estate.

Uicker, John J., Gordon R. Pennock and Joseph E. Shigley. 2003. *Theory of Machines and Mechanisms*. 3rd edn. New York: Oxford University Press.

Vannini, Phillip, Jaigris Hodson, and April Vannini. 2009. 'Toward a Technography of Everyday Life: The Methodological Legacy of James W. Carey's Ecology of Technoculture as Communication'. *Cultural Studies ↔ Critical Methodologies*, 9: 462-76.

Veitch, Henry Newton. 1920. 'Sheffield Plate-II'. *Burlington Magazine*, 37: 18-21, 24-7.

Volta, Alessandro. 1793. 'Account of Some Discoveries Made by M. Galvani, of Bologna'. *Philosophical Transactions*, 83: 10-44.

Vowels, Robin A. 2012. 'The Pilot ACE: From Concept to Reality'. In *Alan Turing's Electronic Brain: The Struggle to Build the ACE, the World's Fastest Computer*, edited by B. Jack Copeland, 223-64. Oxford: Oxford University Press.

Waldrop, M. Mitchell. 2002. *The Dream Machine: J.C.R. Licklider and the Revolution That Made Computing Personal*. London: Penguin.

Walter, W. Grey. 1963. *The Living Brain*. New York: W.W. Norton.

Walter, W. Grey. 1972. 'My Miracle'. *Theoria to Theory*, 6: 38-50.

Wells, H.G. 1995. *The Invisible Man: A Grotesque Romance*. Edited by Macdonald Daly. London: J.M. Dent.

Wertenbaker, G. Peyton. 1929. 'The Chamber of Life'. *Amazing Stories*, October 4: 628-39.

Whitman, Walt. 2004. *The Complete Poems*. Edited by Frances Murphy. London: Penguin.

Wilk, Stephen R. 2013. *How the Ray Gun Got Its Zap: Odd Excursions into Optics*. Oxford: Oxford University Press.

Wilkins, John. 1638. *A Discovery of the World in the Moone*. London: Michael Sparke.

Wilkins, John. 1641. *Mercury, or the Secret and Swifte Messenger; Shewing, How a Man May With Privacy and Speed Communicate His Thoughts To an Absent Friend*. London: John Maynard and Timothy Wilkins.

Wilkins, John. 1646. *Ecclesiastes: Or, a Discourse Concerning the Gift of Preaching as it Falls Under the Rules of Art*. London: for Samuel Gellibrand.

Wilkins, John. 1648. *Mathematicall Magick, or, The Wonders that May be Performed by Mechanicall Geometry*. London: for Samuel Gellibrand.

Wilkins, John. 1651. *Ecclesiastes: Or, a Discourse Concerning the Gift of Preaching as it Falls Under the Rules of Art*. London: for Samuel Gellibrand.

Wilkins, John. 1680. *Mathematical Magick, or, The Wonders that May be Performed by Mechanichal Geometry*. London: for Edward Gellibrand.

Wilkins, John. 1708. *The Mathematical and Philosophical Works of the Right Reverend John Wilkins*. London: for J. Nicholson.

Wolfe, Humbert. 1930. *The Uncelestial City*. London: Victor Gollancz.

Wolfe, Jessica. 2004. *Humanism, Machinery, and Renaissance Literature*. Cambridge: Cambridge University Press.

Woo, Peak. 2010. *Stroboscopy*. San Diego: Plural Publishing.

Woodward, David. 1990. *Feraliminal Lycanthropizer*. San Francisco: Plecid Foundation. http://juniperhills.net/feraliminallycanthropizer.pdf.

Woolgar, Steve. 1998. 'A New Theory of Innovation?' *Prometheus*, 16: 441-52.

Young, James Harvey. 1965. 'Device Quackery in America'. *Bulletin of the History of Medicine*, 39: 154-62.

Index

Abhedânanda, Swâmi 135-6
Abrams, Alfred 151
Adams, George 117
Agrippa, Henry Cornelius 29
Allen, Woody 95-6
Amazon 157-8
androids 24
Archibald, Timothy 88
Archimedes 129
Aristotle 154, 168
Aristophanes 153
Arsić, Branka 95
Artaud, Antonin 137
aura 121
automata 18, 19, 34, 105, 108, 161, 167, 181, 182

Babbage, Charles 13, 26-7, 62
Ball, Philip 149
Baraduc, Hippolyte 135
Barclay, Robert 30
Barrett, Deirdre 58, 59
Bataille, Georges 85-6
Beckett, Samuel 22
Beer, Gillian 145-6
Benjamin, Walter 145
Bennett, Arnold 17-18
Bergson, Henri 16, 88, 152
Berlin, Irving 38
Bernoulli, John 164-5
Bezos, Jeff 157
bicycle 85-8
Bike Smut 86

Birch, John 112
black box 52, 151, 157-8, 176
Blake, William 59
Blaustein, Richard 105
Blavatsky, Helena Petrovna 39, 118
body 17, 22, 24, 29-30, 31, 36, 38, 39, 40, 61, 72-3, 74-5, 91, 99-101, 102, 103-5, 107-17, 134-8, 145-6, 149-51, 170, 173-4
Bohr, Niels 116
Boltzmann. Ludwig 169
Bovet, Richard 29-30
Breakthrough Energy Movement 182
Brentano, Peter 164
Brewster, David 54
Brody, Herb 73
Brooks, Mel 158-9
Browne, Thomas 169-70
Brückner, Burkhart 141
Bühlmann, Vera 34
Bush, Kate 96-97, 131

Caldwell, Roy 95
Calef, Robert 30
Caley, Abraham 35
camera 61
camouflage 148, 153
Campbell, Mary Baine 32-3
Carroll, Lewis 59
Carrouges, Michel 91-2, 94, 175
Carver, Beci 74
Chabris, Christopher 148
Chauliac, Guy de 104-5
cinema 55, 60, 145, 153, 154, 158-9
Clarke, Bruce 44
Collier, Jeremy 28-9
Connolly, Billy 72
comfort 57, 73
Compos, Luis A. 128
Conrad, Joseph, 7, 63-4, 93-4, 126

Corelli, Marie 118-28
Crookes, William 178
cultural technique (Siegert) 104
Curie, Marie 128

de Certeau, Michel 94-5
Deleuze, Gilles 74-6, 77, 79, 91, 94
D'Esperance, Elizabeth 42-3
D'Este, Isabella 154
deus ex machina 153-4
Dircks, Henry 154, 165-6
Disney, Walt 13-14
DOR (Deadly Orgonic Energy) 100
Dreamachine 53-6, 131
dream screen (Lewin) 60
Drebbel, Cornelius van 163
Drown, Ruth 151
drugs 56
Duchamp, Marcel 91, 92-3

Edison, Thomas 130, 132, 182
efficiency 78, 160-2, 165
Einstein, Albert 96
electricity 81, 82, 83, 90, 99-100, 104, 108, 110-19, 127, 158, 181-2
electro-encephalograph 51-2
Ellis, Havelock 84, 85
Empiricus, Sextus 36
energy 16, 35, 41, 69, 74, 83-4, 88, 96, 97, 98-101, 111, 117, 127-8, 132, 134, 139, 148, 151, 164, 168-72, 176-84, 188
engine 12-13
engineering 14-15, 58, 61, 114, 116, 132, 145, 177
Erskine, Ebenezer 35
Espinoza, Tania 22
Euripides 154
explicitation (Sloterdijk) 27
extromission 129

Faithfull, Marianne 55-6
Feraliminal Lycanthropizer 131

Fessenden, Thomas Green 115
flicker 52-5
Fountain, John 46
Franklin, Benjamin 116
Frazer, J.G 107
Freud, Sigmund 15, 21, 46, 61-4, 87, 88, 97, 98-101, 107-9, 121, 127, 139, 141, 144, 147, 155, 178, 185,
Fuchs, Thomas 128, 140-1

Galaxy Quest 35-6
galvanism 107, 110-11
Garrard, George 133
Geiger, Dave 55
Glacier Milk 183
Godwin, Francis 32-3
Goff, Alan 118
Google 51
Graham, James 82-3, 113-14
Gray, Stephen 111
Guattari, Félix 74-6, 77, 79, 91, 94
Guérard, Albert 48
Guibout, Eugène 83-4
Guillemeau, Jacques 105
Gunning, Tom 154
Gysin, Bryon 53, 54-5, 57, 131

Habermas, Jürgen 20
Hamilton, T. Glenn 42
Hamlet 22-3, 126-7
Hansen, Mark 10
Hare, Maurice E. 15
Harness, Cornelius Bennett 114-15
Hearne, Keith 57, 64
Heidegger, Martin 188
Heintzen, L. 134, 138-9
Hervey de Saint-Denys 58-59, 60-1, 63
Hirjak, Dusan 140-1
Hölderlin, Friedrich 188
holism 105-6

homeopathy 106-7
Honnecourt, Villard de 169
Hooke, Robert 144-5
Hovestadt, Ludger 34
Howe, Elias 59
Hughes, Ted 73
Humboldt, Alexander von 107
Humboldt, Wilhem von 168-9
humours 109-10
hypochondria 163

imagineering 13-14
incubation 56, 58-9
induction 82, 117, 127
information 34, 39, 40, 43-5, 60, 167, 169, 170-2, 175-6
ingenuity 12-13, 20, 163
internet 184-5

James, Henry 150
Johnson, Samuel 37
Jonson, Ben 132

Kafka, Franz, 30-1, 36, 49, 91, 92, 93, 149-50
Kant, Immanuel 160
Keely, John Worrell 177-8
Kekulé, August 59
Keller, Evelyn Fox 56
Kempelen, Wolfgang von 155
keratin 72
Kraus, Alfred 141
Krauß, Friedrich 127, 141
Kubrick, Stanley 36

Langelaan, George 43-9
Latour, Bruno 38, 109
laughter 16, 88-9
Lawrence, D.H 18
Lee, Dennis 182
Leonardo da Vinci 162, 163, 169
Leroi-Gourhan, André 21

212 Index

Levesque, Alston 89
Levins, Hoag 89
Lewin, Bertram 60
Lippit, Akira 137
lucid dreaming 57-8, 64

McFadyen, Ian 55, 60
McGill, Donald 86
machines; acoustic 131; anti-gravity 130; attraction (Deleuze and Guattari) 74-5; bachelor (Carrouges) 95, 175; calculating 13, 26, 66-7; celibatory (Carrouges) 75, 91-5; cloud 153-4; cybernetic 161; desiring (Deleuze and Guattari) 74; disintegrating 46-7, 130-1, 132, 138, 177; dream 50-65; experience (Nozick) 76-9; homme-machine (Mettrie) 183; imaginary 7, 9-12, 14, 17, 20-1, 23-4, 27-8, 30, 39, 49, 50, 56, 66, 67, 71, 75, 78-9, 85, 88, 91, 92, 107, 113, 114, 116-17, 123, 125, 126, 129-30, 131-2, 140-2, 148-50, 151, 162-3, 174-5, 176, 182-4, 186-7; influencing (Tausk) 61, 76, 100, 127-8, 134, 137-42; intelligent 186; invisibility 143-59; knitting (Conrad) 7, 63-4, 93-94, 126; medical 103-17; meta- 76, 184; miraculating (Schreber) 75, 77; negentropic 170; overbalancing 175; paranoiac (Deleuze and Guattari) 74-5; perpetual motion 160-85; pleasure 66-102; radiation 118-42; Renaissance 162; repulsion (Deleuze and Guattari) 74; sewing 59, 83-5; sex 79-91; slot 74; time 164, 176, 187; transport 25-49; Ultimate 161, 162; universal (Turing) 184; writing 65, 91-5, 172
MacLeish, Archibald 175
McLuhan, Marshall 174, 184
Maelzel, Johann Nepomuk 155-7, 158
magical thinking 7, 107, 121-2, 144, 151, 178
Maines, Rachel P. 81-2
Marcuse, Herbert 101-2
Martin, William 166
masturbation 72-3, 80-91
mathematics 25, 67
Matthews, James Tilly 127
Maxwell, James Clerk 45, 67, 167, 175-6
Mechanical Turk 155-8
mediation 8, 10, 19, 21-4, 28-9, 32, 34, 38, 50, 57-60, 64-5, 67, 76-80, 102, 104-11, 118, 121-2, 126, 139, 170, 172, 174-5, 184-5,
Mendeleyev, Dmitri 59
metaphor 28-30, 39, 101, 118, 126, 168, 175-6
Mettrie, Julien de la 183

Milton, John 30, 37
Minsky, Marvin 161
Moore, Clara Jessup Bloomfield 177-8
More, Henry 89
Morus, Iwan Rhys 116-17
Moshenska, Joe 37
Müller, Heiner 23

Nelson, Ted 51
Newton, Isaac 26, 111, 129
Not-Machine 23-4, 123, 142
notation 26-7
Novalis (Friedrich von Hardenberg) 7, 93
Nozick, Robert 76-9
number 131-2
nyctograph 59

od 136-7
O'Brien, Flann 86
omnipotence of thoughts (Freud) 121, 124, 144, 178
operation 12, 20, 27, 36, 103-5, 108, 150, 156, 157, 161
orgasm 81, 83, 85, 89, 90-1
orgone 96-7, 98, 100, 107, 131
Orridge, Genesis P. 54

pan-psychism 122
paranoia 74, 99, 132, 134, 137, 139-41, 184
Park, David 129
Parkinson, D.B. 59-60
pathoplasticity 128, 139
Peña, Carolyn Thomas de la 83, 128
Pepper, John Henry 154
Perceval, John 127
Perkins, Elisha 115
Peyrière, Monique 83, 84
phoronomy (Reuleaux) 26
placebo 57
Plato 143-4
play 72-3

Poe, Edgar Allan 155-7
polygraph 15
Pouillet, Thésée 84
Pound, Ezra 180
Pryor, Sean 9
Purdon, James 8, 9
psychotechnography 14-24, 65, 134, 138, 148
Pynchon, Thomas 175-6

Queen 86

Radcliffe Ann 89
radium 119, 128-9, 178
rays 123-4, 129-30
Reddall, John 115
reflexivity 12, 19, 21, 58, 72, 139, 161, 178, 181
Reich, Wilhelm 91, 96-102, 107, 131
Reichenbach, Carl von 137
refrigeration 174
repetition 90
Reuleaux, Franz, 25-7, 127
Richards, I.A. 28
Richet, Charles 41
Ripa, Cesare 163
Ritter, Wilhelm 107
Roth, Vernon 182-3
Roussel, Raymond 157

sadomasochism 125-6
Sartre, Jean-Paul 10, 146-7, 161
Sass, Louis A. 128
Sawday, Jonathan 162, 163
Schaffer, Simon 163-4
Scheerbart, Paul 175
Schreber, Daniel Paul 61, 75, 76, 127, 139, 140, 141
Schwartz, Arturo 93
Sconce, Jeffrey 141
scratching 72
Serres, Michel 50, 133, 137, 170-3

Serviss, Garrett P. 130, 132, 138
Shakespeare, William 11, 22-3, 46, 126-7
Shannon, Claude 43, 161, 170
Sheehan, Nik 55
Shelley, Percy Bysshe 89-90
Sidney, Philip 168
Siegert, Bernhard 104
Simondon, Gilbert 187, 188
Simons, Daniel 148
Sloterdijk, Peter 173
Smith, Kirby Flower 143
Soddy, Frederick 178-80
Sommerville, Ian 53
South Park 86, 184
space craft 31, 36
speed 31, 33, 34, 38-40, 44, 49, 67-8
Spengler, Oswald 146, 166-7
Stead, W.T. 122
Steigerwald, Joan 107
Sterne, Laurence 95
Stiegler, Bernard 21, 23, 187
Stoppard, Tom 47
Sullivan Walter 31

Tausk, Victor 61, 100, 127, 134, 138, 140
technesis 10, 22, 104, 107, 187, 188
technical delusion 127, 132, 134, 139, 141
Technik 10, 105
technique 10
technography 7-10, 21, 24
telautomatics 181, 184
teleplastics 41-3
teleportation 39-49
tension 117
Tesla, Nikola 158, 180-2, 183, 184
thermodynamics 20, 35, 164, 169, 173-6, 179-82, 184; Second Law of 20, 35, 164, 169, 173-5, 176, 183
Tibbits, Herbert 113

time 20, 23, 187
Tinguely, Jean 75, 93
Tobin, Thomas William 154-5
tool 19
transitional object (Winnicott) 22
trick-cyclist 87-8
Trotter, David 9, 40, 132, 134, 137, 139
Trower, Shelley 89
tuning 131-2
Turing, Alan 13, 62, 184
Turner, J.M.W. 133

Van de Graaff, Robert 158
Vaucanson, Jacques de 138
vehicle 28-30
Velasquez, Diego 153
vibration 38, 81-3, 89-90, 95, 130, 132-3, 136, 177, 178
Vieira-Schmidt, Vanda 139
visualisation 11, 137, 145, 158-9
Vitruvius (Marcus Vitruvius Pollio) 153
Volta, Alessandro 110

Waldrop, M. Mitchell 51
Walter, W. Grey 51-3
Wells, H.G. 130, 136, 147-8, 151-3, 176
Wertenbaker, Green Peyton 79-80
Wesley, John 115-16
Whale, James 147, 158-9
Whitman, Walt 104
Whitney, John 60
Wilkins, John 33-4, 37, 39, 67-71
Winnicott, Donald 22
Wolfe, Humbert 87-8
Wolfe, Jessica 162
Womersley, John 13
Woodward, David 131
Woolf, Virginia 171

X-rays 128, 129, 134-7

www.ingramcontent.com/pod-product-compliance
Lightning Source LLC
Chambersburg PA
CBHW030854170426
43193CB00009BA/611